THE PHYSICS OF BEAMS
ANDREW SESSLER SYMPOSIUM

AIP CONFERENCE PROCEEDINGS 351

THE PHYSICS OF BEAMS

ANDREW SESSLER SYMPOSIUM

LOS ANGELES, CA   DECEMBER 1993

EDITOR: WILLIAM A. BARLETTA
LAWRENCE BERKELEY
LABORATORY

American Institute of Physics   Woodbury, New York

Authorization to photocopy items for internal or personal use, beyond the free copying permitted under the 1978 U.S. Copyright Law (see statement below), is granted by the American Institute of Physics for users registered with the Copyright Clearance Center (CCC) Transactional Reporting Service, provided that the base fee of $6.00 per copy is paid directly to CCC, 222 Rosewood Drive, Danvers, MA 01923. For those organizations that have been granted a photocopy license by CCC, a separate system of payment has been arranged. The fee code for users of the Transactional Reporting Service is: 1-56396-376-0/96 /$6.00.

© 1996 American Institute of Physics.

Individual readers of this volume and nonprofit libraries, acting for them, are permitted to make fair use of the material in it, such as copying an article for use in teaching or research. Permission is granted to quote from this volume in scientific work with the customary acknowledgment of the source. To reprint a figure, table, or other excerpt requires the consent of one of the original authors and notification to AIP. Republication or systematic or multiple reproduction of any material in this volume is permitted only under license from AIP. Address inquiries to AIP, Office of Rights and Permissions, 500 Sunnyside Boulevard, Woodbury, NY 11797-2999; phone 516-576-2268; fax: 516-576-2499; e-mail: rights@aip.org.

L.C. Catalog Card No. 95-80497
ISBN 1-56396-376-0
DOE CONF-931-253

Printed in the United States of America.

CONTENTS

Preface vii
Linear Colliders: The Last Ten Years and the Next Ten Years 1
    R. H. Siemann
Manipulating Charged Particle Beams and Light by Means of Plasma 17
    J. M. Dawson
Excitation of Accelerating Wakefields in Inhomogeneous Plasmas 24
    G. Shvets and J. S. Wurtele
Collective Instabilities in Accelerator and Storage Rings 49
    C. Pellegrini
Maximizing the Luminosity of ELOISATRON, A Hadron Supercollider
at 100 TeV Per Beam 56
    W. A. Barletta
Possible Applications of Plasma Lens in High Energy Physics 68
    P. Chen
MURA Days 79
    K. R. Symon
Andrew Sessler's LBL Directorship 96
    E. K. Hyde
Andrew Marienhoff Sessler, *Physicist—A Student's Guide to the Couplings
of a Beam Physicist with His Environment* 101
    D. H. Whittum
Andy Sessler: A Physicist for All Seasons 120
    M. Pripstein
Curriculum Vitae 125
    A. M. Sessler
Author Index 149

Preface

In December 1993, more than one hundred of Andrew Sessler's friends and colleagues gathered in Berkeley for a symposium celebrating his 65th birthday and honoring over forty years of major scientific contributions to accelerator and beam physics. The technical topics of his research have ranged widely: accelerator and collider design, beam instabilities, beam interactions with plasmas, and techniques for producing coherent radiation (such as free electron lasers).

Andy's contributions to our community have been scientific, institutional, and personal. Throughout the world hundreds of his collaborators have shared research explorations and numerous animated discussions. He has been a mentor both to students and to experienced physicists. Those of us at the Lawrence Berkeley National Laboratory know him as our former Director, who led the Laboratory through a difficult transition when the Bevatron ceased to be a focus for experimental high energy physics. Many more scientists, especially in the states of the former Soviet Union, have had their lives touched by Andy's tireless devotion to promoting human rights and open scientific inquiry.

This book represents a selection of the oral presentations given at the Sessler Symposium plus some added contributions all reflecting Andy's technical interests. Once, in the course of our preparing a paper about FEL physics, Andy told me that he liked doing "impressionistic physics" leaving to his collaborators the task of sharpening the contours, distinctions, and quantitative implications. This volume likewise presents his friends' impressionist view of Andy's range of interests and activities as a researcher, as a scientific leader and mentor, and as a person.

In choosing topics for his research Andy frequently comments that "it would be fun to look at ..." To Andy "fun" neither demeans the research topic nor reduces its exploration to self-indulgence. To the contrary, it emphasizes that pushing out in new directions is intellectually engaging and self-motivating, that many unforeseen paths and options will open from successful work. All the contributors to this volume are indebted to Andy for opening the opportunity for so much "fun."

William A. Barletta
Accelerator and Fusion Research Division
E. O. Lawrence Berkeley National Laboratory
July, 1995

LINEAR COLLIDERS: THE LAST TEN YEARS AND THE NEXT TEN YEARS

Robert H. Siemann[*]
Stanford Linear Accelerator Center, Stanford University, Stanford, CA 94309

INTRODUCTION

Some of the most important discoveries and systematic studies in elementary particle physics have been made at electron-positron colliders. These include the discoveries and measurements of the properties of the C-quark and τ-lepton, the studies of B-mesons and gluons, the measurement of the number of light neutrinos, and precision measurements of electroweak parameters. These colliders are such powerful instruments because of the unique center of mass energy and initial quantum numbers, $J^{PC} = 1^{--}$, of e^+e^- annihilation, and backgrounds that are beam-related rather than being an unavoidable part of the total cross section.

Storage rings are limited in center-of-mass energy, E_{CM}, by synchrotron radiation. The synchrotron radiation energy loss per turn is

$$U_0 = \frac{4\pi r_e}{3} mc^2 \frac{\gamma^4}{\rho} \qquad (1)$$

where γ is the beam energy in units of rest energy, mc^2, r_e is the classical electron radius, and ρ is the bending radius. The luminosity of a storage ring operating at the beam-beam limit is directly proportional to the total current, I_T, and beam energy

$$L = \xi \frac{I_T}{e} \frac{\gamma}{r_e \beta_y^*}. \qquad (2)$$

In this equation, ξ is the beam-beam tune shift and β_y^* is the vertical beta-function at the collision point. Non-resonant cross sections fall as $1/\gamma^2$, and the synchrotron radiation power, $P_{SR} = I_T U_0$, must increase as γ^5 for a constant event rate.

LEP is the largest storage ring in the world; some of its parameters are given in Table I. These numbers together with the steep energy dependences of U_0 and P_{SR} lead to the conclusion that the size and cost of a storage ring collider with a center-of-mass energy much greater than that of LEP would be astronomical! Linear colliders avoid this energy limit by not bending the beams, and they extend the potential energy of e^+e^- collisions.

The cross section for producing μ-pairs is

$$\sigma(e^+e^- \to \mu^+\mu^-) = \frac{\pi \alpha^2}{3(mc^2)^2} \frac{1}{\gamma^2}. \qquad (3)$$

This cross section is 87 fb at $E_{CM} = 1$ TeV, and $L = 10^{33}$ cm^{-2}s^{-1} at that energy would give 7.5 μ-pairs per day. All of the cross sections for producing point like particles are proportional to the μ-pair cross section, and luminosity in the $10^{33} - 10^{34}$ cm^{-2}s^{-1} range is needed for the physics at roughly 1 TeV. This demanding requirement dominates high energy linear collider design.

---

[*] Work supported by the Department of Energy, contract DE-AC03-76SF00515.

© 1996 American Institute of Physics

Table I LEP Parameters [1, 2]

Parameter	LEP I	LEP 200
Circumference	\multicolumn{2}{c}{26,659 m}	
Bending Radius, ρ	3,096 m	
Total Current, I_T	3.0 mA	
Beam Energy, E	55 GeV	95 GeV
Energy Loss/Turn, U_0	260 MeV/turn	2.31 GeV/turn
Synch. Rad. Power, P_{SR}	1.56 MW	13.9 MW
Peak RF Voltage	360 MV	2.7 GV
Nominal Luminosity	$1.7 \times 10^{31} \text{cm}^{-2}\text{s}^{-1}$	$2.7 \times 10^{31} \text{cm}^{-2}\text{s}^{-1}$

Linear colliders were first proposed by Maury Tigner in 1965,[3] but the interest in them heated up in the early 1980's when the implications of the energy limitation from synchrotron radiation was first appreciated in a concrete way. Until then it was easier to increase the size of storage rings rather than face the energy limit head-on. LEP made it clear that this approach had reached its end, and linear colliders would have to be developed to increase the energy of e^+e^- collisions.

The early 1980's was a period of great enthusiasm about linear colliders. The SLC had been approved and was under construction. A quick turn-on and a first year luminosity of $6 \times 10^{29} \text{cm}^{-2}\text{s}^{-1}$ were projected.[4] In addition, people talked loosely about a linear collider that could do "SSC equivalent" physics at a fraction of the cost. These projections were beyond what could be supported by reasonable expectations, but the enthusiasm was critical because it set into motion a series of actions that are forming the three thrusts in linear collider development. These were:

1.- The construction, commissioning, and development of the SLC that has established the viability of linear colliders;
2.- The research and technology development aimed at an $E_{CM} = 0.5$ TeV linear collider that has opened up a new energy range for e^+e^- collisions;
3.- The establishment of the field of advanced accelerators and new particle acceleration techniques that hold out the possibility of e^+e^- collisions determining the energy frontier of particle physics in the future.

Traditionally that energy frontier has belonged to the hadron colliders where it hasn't been necessary to deal with synchrotron radiation. That has more than compensated for the constituent center-of-mass energy being lower than the beam center-of-mass energy. There are technically sound ways to reach constituent center-of-mass energies of several TeV with hadron colliders, but they will have reached an energy where storage rings become impractical after completion of the LHC. An innovation as novel as the linear collider for e^+e^- collisions would be needed to reach significantly higher energies, and e^+e^- linear colliders could determine the energy frontier in the future. That will require another step beyond the $E_{CM} = 0.5$ TeV collider that is the present focus of the linear collider community, and it is likely to require success with some of the directions being pursued by the advanced accelerator community.

The energy frontier is uncertain and insecure. Despite its outstanding science and technical merit the SSC failed because of a combination of politics, cost, and economic climate in the United States. The LHC has not been approved yet, and its construction is not assured. Political and economic issues similar to those for the SSC and LHC are certain to be in the future for linear colliders. To face them one needs to be optimistic that outstanding, large scientific projects will be supported and that methods for that support will be developed through the experience with the LHC, large fusion and space research projects, and the informal international collaborations developing linear collider designs .

THE SLC

The SLC was intended as a prototype linear collider and as an accelerator for particle physics. These are compatible because of the large cross section for producing Z's, $\sigma(e^+e^- \to Z) \approx 30$ nb. Demonstration of particulars such as small collision spots or high intensities were crucial steps in the development of the SLC, but, by themselves, they are not enough to prove the viability of linear colliders. That depends on sustained operation with good luminosity and low backgrounds with the demanding conditions set by a particle physics experiment.

The SLC had a long, difficult commissioning. It was the first of a new type of accelerator, and despite the best efforts of the designers, the difficulties of operating it were not appreciated beforehand. Many parameters including peak current, transverse emittances, and beam sizes were being pushed into new regimes simultaneously. Linear colliders are like hadron colliders in one way - any mistake made upstream is remembered and emittance preservation, β matching, dispersion matching, etc. are extremely important. This makes it difficult to achieve performance breakthroughs in many areas simultaneously. Problems in downstream areas such as the final focus can't even be seen until upstream areas are performing reasonably, and collision performance cannot be used to diagnose problems upstream until the downstream is performing reasonably. In many cases new diagnostics and techniques had to be invented and proved to make progress.

The complexity of the SLC is in a new regime also. The engineering standards for performance and reliability are stringent for a linear collider because there are a large number of components and few of them can fail or be operating out of specification without impacting performance. This is another factor that made the SLC commissioning difficult. New levels of equipment and beam diagnostics and control were needed. It took time for this to be realized and for these to be developed. Now they are part of everyday operation.

SLC commissioning is over, and the SLC is running as an accelerator for particle physics and as a prototype linear collider. The day-to-day operation and luminosity improvement programs are shaping many of the ideas about future linear colliders.

Table II and Figures 1, 2, and 4 summarize performance. There has been steady improvement in the average luminosity which reached a peak of 3.5×10^{29} cm^{-2}s^{-1} early in the summer of 1993. The increase from 1992 to 1993 came from the synergy between SLC operation and work on future colliders which are based on flat beams, $\sigma_x \gg \sigma_y$, to minimize backgrounds. Emittance preservation and focusing of flat beams with high order optical corrections need to be verified experimentally as part of the development of future colliders. The Final Focus Test Beam (FFTB) collaboration was formed to develop and test a prototype next generation final focus. They need flat beams for those tests, and in the process of delivering those beams it was found that the alignment, feedback and orbit bump techniques developed at the SLC performed better than expected. Invariant emittances well below the design value

Table II Typical SLC Parameters for the 1993 Run

Parameter		Value	
Energy	at End of Linac	46.5 GeV	
	at Interaction Point	45.59 GeV	($m_Z/2$)
Intensity	in Damping Rings	$3.0 - 3.3 \times 10^{10}$	
	at Interaction Point	$2.8 - 3.0 \times 10^{10}$	
Polarization		0.60 - 0.64	
Invariant Emittances	at End of RTL	$3 - 4 \times 10^{-5}$ m	horizontal
		$0.3 - 0.4 \times 10^{-5}$ m	vertical
	at End of Linac	$4 - 5 \times 10^{-5}$ m	horizontal
		$0.5 - 0.9 \times 10^{-5}$ m	vertical
Beam Size	at Interaction Point	2.6 µm	horizontal
		0.8 µm	vertical
RMS Bunch Length	at Interaction Point	1.0 mm	
RMS Energy Spread		0.3%	
Repetition Rate		120 Hz	
Luminosity	(see Figure 1)	$0.2 - 0.35 \times 10^{30}$ cm^{-2}s^{-1}	

of 3×10^{-5} m could be transported down the linac at an intensity of 3×10^{10} particles per bunch.[5] These results showed that it was possible to use flat beams in the SLC. The SLC arcs prevent running with truly flat beams but elliptical beams are possible, and the reduction in beam area (Figure 2) in 1993 came from that.[6]

With flat beams and the current SLC final focus the minimum vertical beam size is about 0.8 µm (Figure 3) and is dominated by a single third-order aberration, the

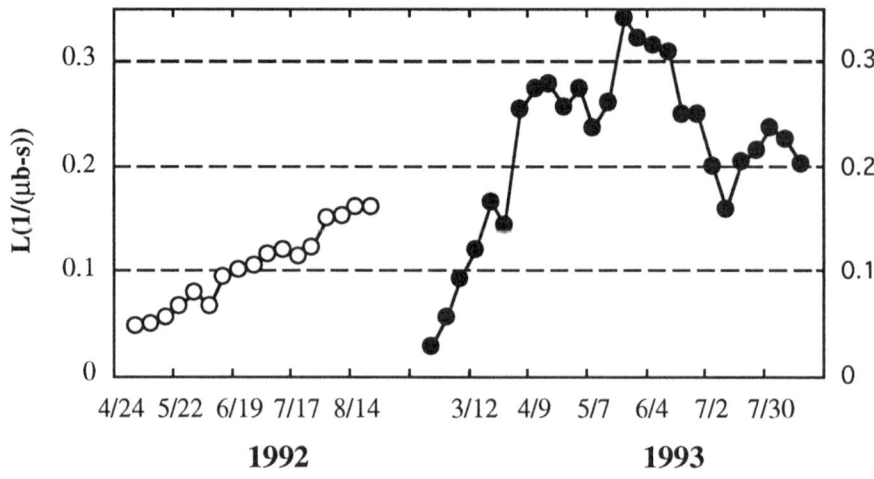

Figure 1. The average SLC luminosity (in units of 10^{30}cm^{-2}s^{-1}) in 1992 and 1993.

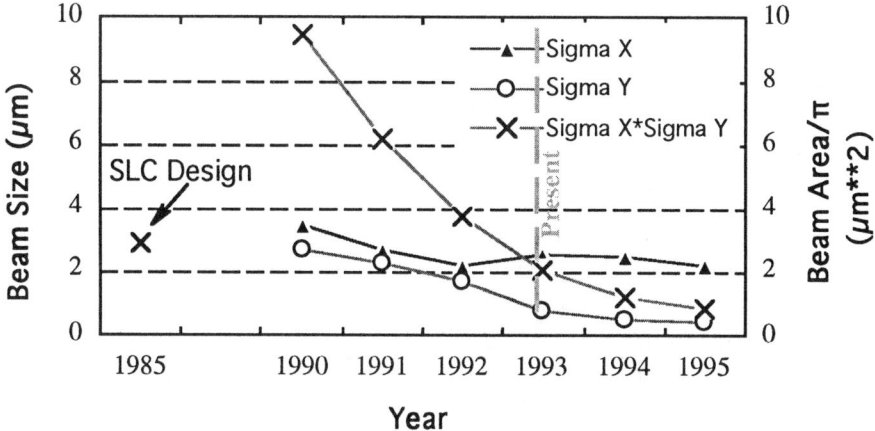

Figure 2. The transverse beam sizes and beam area at the SLC collision point. Past performance and future projections are shown.

$y'^2 \delta^2$ term in the Hamiltonian (U_{3466} in TRANSPORT notation). This aberration can be controlled with an additional quadrupole in the chromatic correction section (CCS). A major SLC upgrade for 1994 is the installation of that quadrupole, some sextupoles to correct geometric aberrations in the final triplet, and additional optics and diagnostics to aid final focus tuning.[7]

The SLC has relied on developments for future colliders in making the performance improvement in 1993 and designing the 1994 upgrade. SLC operation in 1994 will address a vital concern for future colliders - can a highly corrected optical system be diagnosed and tuned fast enough and precisely enough to accommodate the continual changes in the incoming beam from the linac? This development is just one example of the importance of the SLC for future linear colliders.

The luminosity fell after the peak in the early summer due in part to a heat wave and interruptions to steady running. However, these factors alone cannot explain the reduced luminosity; possible causes include shifts in alignment due to the changing water table and deterioration of the vacuum in a section of the accelerator. These type of problems occur in any accelerator, and learning to diagnose and fix them quickly is key to good integrated luminosity. Some of the new final focus diagnostics were motivated by not being able to understand the luminosity decrease, and if they prove successful at quickly identifying problems, SLC operation will have added even more to the specifications of the diagnostics needed in any future collider.

The integrated luminosity is shown in Figure 4. The SLC has been operating with between 60% and 80% uptime, and the SLD experiment accumulated over 10,000 $e^+e^- \to Z$ events during the 1992 run when the electron beam polarization was P ~ 22%. The first measurement of the left right asymmetry in Z boson production, A_{LR}, was published based on those data.[8] The statistical uncertainty of that measurement is proportional to $1/(P\sqrt{N_{ev}})$ where N_{ev} is the number of events. The polarization and number of events were increased significantly in 1993 to 62% and 50,000, respectively, and there should be a new, precise measurement of A_{LR} and the Weinberg Angle based on those data soon. The SLC is meeting its goal of being an accelerator for high energy experiments.

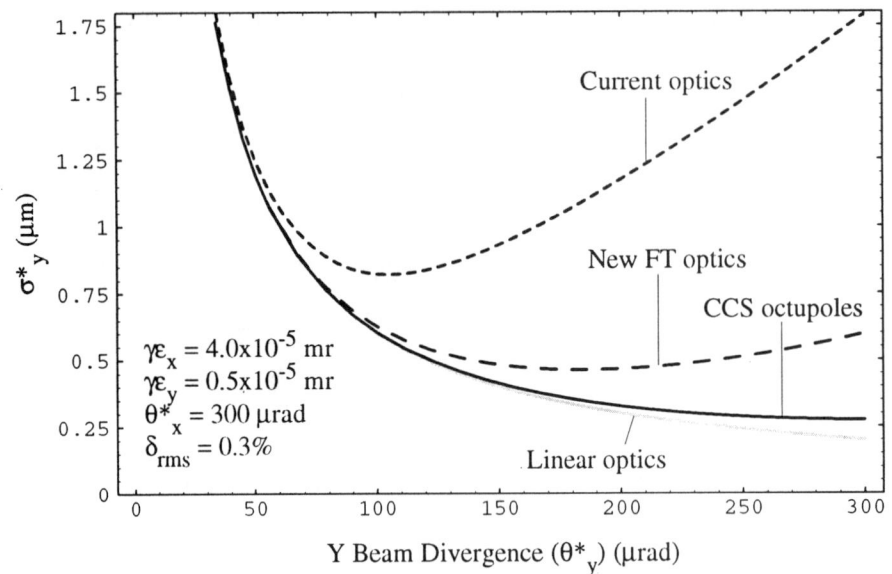

Figure 3. SLC final focus performance for the optics used in 1993 ("Current optics"), the upgrade being installed for 1994 ("New FT optics"), and for a possible future improvement ("CCS octupoles").[7]

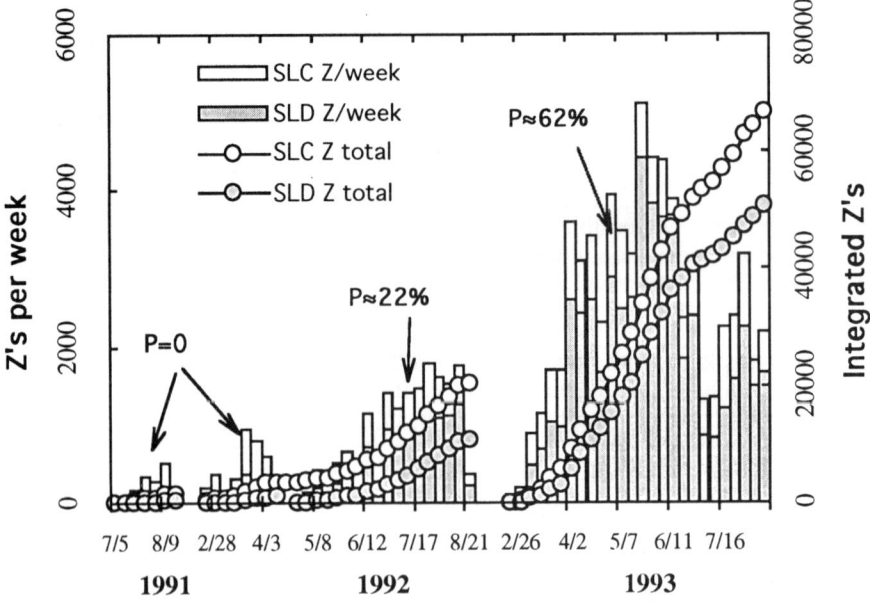

Figure 4. SLC performance for the past three years. The SLC numbers are based on beam current and spot size measurements from beam-beam deflections. The SLD numbers are the number of Z's recorded on tape.

The second major improvement nearing completion is replacement of the damping ring vacuum chamber. The damping ring bunch length increases due to potential well distortion starting at about 2×10^{10} particles per bunch, and the microwave instability threshold is 3×10^{10} per bunch. The transverse emittances are increased by potential well distortion because of the following. A longitudinal phase space rotation is performed in the Ring-To-Linac (RTL) transfer line interchanging energy spread and bunch length, so potential well distortion increases the energy spread in the RTL. The RTL must be chromatically corrected to preserve the transverse emittances,[9] and that correction is not as good and becomes more sensitive and difficult to maintain against drifts as the energy spread increases.

The microwave instability in the damping rings is the present SLC intensity limit. It manifests itself as a relaxation oscillation.[10] The beam radiation damps until the peak current exceeds the instability threshold; then the longitudinal phase space blows-up rapidly, in ~5 synchrotron oscillations, to a peak current that is below threshold. Radiation damping starts the cycle over again. The synchronous phase shifts during this oscillation because of the bunch length dependence of the higher mode losses. Shifts in the synchronous phase and bunch length affect the bunch rotation in the RTL and the phase of injection into the linac. Bunches extracted when the longitudinal phase space is changing rapidly are handled particularly poorly and can be so far off in energy that the backgrounds they create trip off the SLD detector. The only practical way to avoid this has been to restrict the maximum intensity to near the microwave instability threshold.

The present damping ring impedance is dominated by masks protecting bellows from synchrotron radiation, transitions between different chamber geometries, and distributed ion pump slots. This impedance is being reduced by a factor of five through a combination of precision manufacturing and magnet alignment to reduce the number of bellows and numerically controlled machining to make the transitions more gradual and the ion pump slots narrower. With this new vacuum chamber the potential well distortion and microwave instability thresholds will be well above the charge that the linac can accelerate to 46.5 GeV, the energy needed for collisions at $E_{CM} = m_Z$.

It is projected that the final focus and damping ring upgrades will bring the SLC luminosity to $L > 10^{30} cm^{-2} s^{-1}$ and the event rate to over 10,000 Z's per week by decreasing the vertical size to $\sigma_y \leq 0.5$ μm and increasing the intensity at the collision point to $3.5 - 4\times 10^{10}$. It is difficult to predict parameters and performance precisely because the upgrades will remove the present spot size and intensity limits and there isn't any experience yet to know the next limits. There should be substantial disruption that should increase the luminosity by roughly 30%.

A wide range of knowledge has been learned from the SLC. It can be characterized as:
1.- The development of particular components or systems that were pushing the state-of-the-art and are now the standard for comparison and the base for the next developments. Examples are the 60 MW klystrons, the positron target and capture system, and the polarized electron gun.
2.- Techniques that have proven their value in SLC operation and have been incorporated as central features of the next generation collider. These include beam based alignment and optics diagnostics, techniques for emittance preservation, and adaptive feedback systems.
3.- Beam dynamics including experiments in emittance preservation, polarization control, and damping ring beam instabilities. The anticipated observation of disruption next year will be the first experience with beam-beam effects in linear colliders.

4.- Design philosophies for future colliders. Linear colliders are complex, and that complexity is bound to increase with increasing energy. The interdependence of the SLC accelerator systems made commissioning difficult, and it still shows up today in subtle ways such as trips of the SLD drift chamber high voltage being the first, most evident manifestation of the damping ring microwave instability. Reliability and quality assurance (QA) have unfortunate, bureaucratic connotations, but specifying reliability and treating it on a par with performance and cost would have decreased the SLC commissioning time significantly. Finally, thorough diagnostics of the beam and equipment has a handsome payoff. The SLC control system is designed to routinely measure and log an enormous amount of data such as beam emittances, power supply control and readback signals, and temperatures in all parts of the accelerator. These data are invaluable for finding and fixing problems which often are not evident when they first occur.

$$E_{CM} = 0.5 \text{ TEV}[11]$$

Linear collider design and development have become focused on $E_{CM} = 0.5$ TeV and $L \sim 5 \times 10^{33} \text{cm}^{-2}\text{sec}^{-1}$. The consensus on these general parameters has come about because they combine technical feasibility with substantial particle physics including studies of the top quark, possible studies of electroweak symmetry breaking phenomenon, and large enough increases in center-of-mass energy and luminosity to reveal the totally new and unexpected. A recent ICFA Seminar strongly endorsed a 0.5 TeV linear collider as the next natural step for high energy physics after the LHC and as an important opportunity for international collaboration.[12]

There are diverse approaches to meeting these general objectives. The diversity arises from different judgments about the ease of developing new and improving existing technology, costs, extension to higher energies, experimental backgrounds, center-of-mass energy spectrum, tolerances, and beam power.

Selected parameters are given in Table III which is based on a compilation made by G. Loew at the LC-93 Conference and is reproduced with his permission.[13] The colliders described in that table are:

1.- TESLA (being developed by an international collaboration) which is based on superconducting RF. All the others would use room temperature RF.
2.- SBLC (an international collaboration centered at DESY/Darmstadt) which uses S-band (3 GHz) RF where there is extensive operating experience.
3.- NLC (SLAC) which uses higher frequency X-band (11.4 GHz) RF in a modulator-klystron-accelerator configuration similar to S-band linacs.
4.- JLC-I (KEK) which has three frequency options, S-band, C-band (5.7 GHz), and X-band. Multiple bunches are accelerated in each RF pulse as they are in TESLA, SBLC, and NLC.
5.- VLEPP (INP) which employs a single high intensity bunch rather than multiple bunches.
6.- CLIC (CERN) which is a "two-beam" accelerator with klystrons replaced by an RF power source based on a high-current, low-energy beam traveling parallel to the high energy beam.

The AC power is large for any of the colliders, and energy efficiency is important. One way to achieve good efficiency is by accelerating multiple beam bunches per RF pulse.

For example, in the SBLC a 150 MW, 2.8 μsec long RF pulse powers two 6 m long sections to a gradient of 17 MV/m. The beam has 125 bunches with 2.9×10^{10}

Table III Selected Linear Collider Parameters for $E_{CM} = 0.5$ TeV
(G. Loew, LC93)[13]

Parameter	TESLA	SBLC	JLC-I (S)	JLC-I (C)	JLC-I (X)	NLC	VLEPP	CLIC
L (10^{33}cm^{-2}s^{-1})	7	4	4	7	6	8	15	2 - 9
RF Freq (GHz)	1.3	3.0	2.8	5.7	11.4	11.4	14	30
Rep Rate (Hz)	10	50	50	100	150	180	300	1700
Bunches per RF pulse	800	125	55	72	90	90	1	1 - 4
N (10^{10})	5.15	2.9	1.30	1.0	0.63	0.65	20.	0.6
BPM Precision (μm)†	10.	10.	NA	NA	1.	1.	0.1	0.1
$\gamma\epsilon_x/\gamma\epsilon_y$ (10^{-8}m)	2000/100	1000/50	330/4.5	330/4.5	330/4.5	500/5	2000/7.5	180/20
β_x^*/β_y^* (mm)	25/2	22/0.8	10/0.1	10/0.1	10/0.1	10/0.1	100/0.1	2.2/.16
σ_{x0}/σ_{y0} (nm)	1000/64	670/28	300/3	260/3	260/3	300/3	2000/4	90/8
σ_L (μm)	1000	500	80	80	67	100	750	170
IP Crossing Angle (mrad)	0	3	7.3	8	7.2	3	--	1
Y	0.029	0.055	0.24	0.21	0.16	0.096	0.074	0.35
H_D	2.3	1.6	1.6	1.5	1.7	1.4	1.3	3.3
δ_B	0.03	0.03	0.10	0.08	0.05	0.03	0.13	0.36
n_γ	2.7	2.0	1.6	1.4	1.0	0.9	5.0	4.7
Loaded Grad. (MV/m)	25	17	19	33	31	38	96	78 - 73
Two Linac Length (km)	20	29.4	28	16.7	17.7	14	6.4	6.6
Section Length (m)	1.04	6	3.6	2	1.3	1.8	1.01	0.273
Number of Sections	19232	4900	7776	8360	13600	7778	5200	24000
Number of Klystrons	1202	2450	1944	4180	3400	1945	1300	2
Klystron Peak Power (MW)	3.25	150	85	45	70	94	150	700
Klystron Pulse Length (μs)	1300	2.8	4.5	3.6	0.84	1.5	0.7	0.011
Pulse Length to Section (μs)	1300	2.8	1.2	0.6	0.21	0.25	0.11	0.011
Pulse Compression Gain	--	--	2.4	4.2	3.2	4	4.22	--
a/λ (input/output cavity)	0.15	.15/.11	0.13	.16/.12	.24/.14	.22/.15	.14	.2
P_B (MW)	16.5	7.3	1.4	2.9	3.4	4.2	2.4	.4 - 1.6
AC Power (MW)	137	114	106	193	86	141	91	175
$2P_B/P_{AC}$	0.24	0.13	0.03	0.04	0.09	0.06	0.05	0.02

† Addition to G. Loew's compilation (from Ref. 14).

particles per bunch spaced 16 nsec apart. The RF pulse has 420 J of energy; a single bunch extracts 0.95 J from the accelerator RF fields, and the bunch train extracts a total of 118 J leading to an efficiency, η_B, for converting RF to beam energy of η_B = 0.28. If only a single bunch was accelerated, the RF pulse could be shortened to 1 μsec, the accelerator filling time, but the efficiency would be low, η_B = 0.006. A major advantage of multiple bunches is that the cost of filling the accelerator with RF energy has been amortized over a large number of bunches.

Multiple bunches have implications for both the fundamental and higher modes. The energy spread of the beam must be small to minimize emittance blow-up from dispersive effects in the linac and to minimize chromatic aberrations in the final focus. The bunch train lengths are comparable to filling times, and the accelerator structure must be prefilled and the RF amplitude ramped so that each bunch gains the same energy.[15]

The bunches are closely spaced, and they interact through higher modes. The transverse modes can cause emittance blow-up that is in addition to that from the short range transverse wakefield. The interaction between bunches must be reduced by damping higher order modes or by detuning, varying cell dimensions to spread mode frequencies, leading to destructive interference between the deflections from different cells.[16] Detuning and damping may have to be combined to get adequate reduction of the long range wakefields.

VLEPP has a single, large bunch, 2×10^{11} particles, and that results in η_B = 0.12. The large bunch and relatively high RF frequency impose stringent tolerances on the linac for emittance preservation and require a novel final focus, the traveling focus, where a head-tail energy shift is introduced to shift the focal point during the collision and prevent enormous disruption. CLIC has parameters for between one and four bunches, and studies of energy compensation and transverse modes for four bunches are in progress.[17]

Present day, conventional linacs are modular with each module consisting of a modulator, klystron, possibly an RF pulse compression system, and, finally, one or more accelerator sections powered in parallel. The modulator converts AC power to high voltage, pulsed power. Most use a low voltage, lumped element transmission line for energy storage, thyratrons as switches, and a pulse transformer to step-up the output voltage. SLAC modulators are typical and are roughly 75% efficient.[18] A substantial fraction of the inefficiency comes from the rise- and fall-times of the pulse transformer. Improving modulator efficiency would be significant. A capacitor bank and high voltage switch tube rather than a pulse transformer was being considered for the SBLC, but has been given up for lack of an appropriate switch tube. A DC high voltage supply and girded klystron is being developed for VLEPP.

A short, high power RF pulse is the ideal for high frequencies because short sections and high group velocities are favored by efficiency and wakefields. The input power must be multiplied by $\tau^2/(1 - e^{-\tau})^2$ for the same average accelerating gradient ; $\tau \propto \zeta/(\lambda^{1.5}\beta_g)$ where ζ is the section length, β_g is the (normalized) group velocity, and λ is the RF wavelength.[19] The wavelength dependence comes from the skin effect. The maximum transverse wakefield behaves as $1/(a^3(\lambda/a)^{.8})$ where a is the radius of the waveguide iris.[20] Increasing l/a reduces the wakefield with the side effect of raising the group velocity.[19]

It is impractical to generate short RF pulses directly. Modulator efficiency would be poor because pulse rise-and fall-times would be a large fraction of the pulse and klystron peak power would be enormous. Pulse compression[21] which raises the peak power while shortening the RF pulse is used for matching klystron capabilities to an optimum accelerator configuration and is a feature of the high RF frequency colliders.

TESLA has unique power source requirements. The high Q and long pulse length reduce the peak power to 3.25 MW, but the modulator must be capable of delivering that power for over a millisecond.

All except CLIC have a large number of klystrons. CLIC is a two-beam accelerator which replaces all of this with a single, low-energy beam traveling parallel to the high energy beam. This low-energy beam has a time structure appropriate for generating 30 GHz RF. It is accelerated by a superconducting RF system, and energy is extracted with transfer structures spaced roughly 1.5 m apart.

The vertical invariant emittances, $\gamma\epsilon_y$, are small, and emittance preservation during acceleration is an important consideration. Emittance growth caused by the combination of injection jitter and wakefields must be controlled by tight tolerances on injection elements and BNS damping.[22] Those tolerances range from about 1 μm for NLC and JLC-I(X) to about 10 μm for the S-band accelerators and TESLA.[14]

Misalignments in the main linac cause emittance growth through wakefields and dispersion. With straight one-to-one orbit correction, i. e. steering to the middle of beam position monitors, there would be extremely tight tolerances on accelerator, quadrupole, and beam position monitor alignment. As examples, those tolerances would be about 10 μm for SBLC and half that for NLC.

Beam-based orbit correction procedures, where optical elements are varied and orbit changes measured, relieve these tolerances substantially.[14] The strengths of all the quadrupoles are increased, or decreased, in Dispersion Free (DF) steering to measure momentum dependence of the central trajectory; then, the orbit is corrected to minimize the dispersion. The strengths of focusing quadrupoles are reduced while those of defocusing quadrupoles are raised to approximate the defocusing effect of wakefields in Wakefield Free (WF) steering. Wakefield free steering requires good local alignment between quadrupoles and accelerator sections. Since these procedures depend on measuring orbit changes, the beam position monitors (BPM's) must be precise. Estimates of the required precisions are included in Table III and range from 0.1 μm for CLIC and VLEPP to 10 μm for SBLC and TESLA.[14]

The beams are flat at the interaction point to minimize backgrounds (see below) with $\gamma\epsilon_x \gg \gamma\epsilon_y$ and $\beta_x^* \gg \beta_y^* > \sigma_L$ (for all but VLEPP with its traveling focus) where σ_L is the bunch length. The vertical dimension is the most demanding with the vertical sizes before disruption ranging from 64 nm (TESLA) to 3 nm (JLC, NLC).

The vertical spot sizes quoted are the first order sizes, $\sigma_{y0} = (\beta_y^* \epsilon_y)^{1/2}$, and up to third order geometric and chromatic aberrations must be corrected to reach those sizes. This is done by using dipoles to introduce dispersion in a region with sextupoles separated by a -I transformation. Synchrotron radiation losses in the chromatic correction section and in the final quadrupoles introduce important aberrations.

There are extremely tight pulse-to-pulse jitter tolerances. For all but the final doublet those tolerances are about $10\sigma_y$ while for the final doublet they are roughly σ_y.[23] The Final Focus Test Beam (FFTB) at SLAC will test many of the techniques for reducing aberrations to the required level and will provide a test bed for studying and specifying jitter tolerances.

The beams cross at an angle. This avoids unwanted collisions for colliders with closely spaced bunches, and it allows the channel for focusing the incoming beam to be independent of the channel for the exiting disrupted beam. Crab crossing,[24] tilting the bunches with an RF deflector, prevents luminosity loss due to incomplete overlap.

The luminosity is given by

$$L = \frac{N^2 f_c}{4\pi\sigma_{x0}\sigma_{y0}} H_D = \frac{N^2 f_c}{4\pi\sigma_x\sigma_y} \; ; \qquad (4)$$

N is the number of particles/bunch and f_c is the collision frequency. Focusing during the collision, disruption, is accounted for by an enhancement factor, H_D, in the left-hand expression where the beams sizes without disruption are used, and by using the disrupted beam sizes in the right-hand expression.

The electromagnetic fields at the collision point are parametrized by[25]

$$Y = \frac{5 r_e^2}{6\alpha} \frac{\gamma N}{\sigma_L(\sigma_x + \sigma_y)} . \qquad (5)$$

Field enhancement due to disruption is accounted for approximately by using the disrupted sizes. This increases Y for TESLA, SBLC, and CLIC because the horizontal size is reduced about 50% by disruption in those cases. The mean energy beamstrahlung energy loss, $\delta_B \propto Y^2$, and backgrounds from beamstrahlung, e^+e^- pairs, and hadronic events depend on Y. When $Y \ll 1$ and $\sigma_x \gg \sigma_y$, the mean number of beamstrahlung photons per incident particle is[25]

$$n_\gamma \cong \frac{5\alpha^2 \sigma_L}{2 r_e \gamma} Y \cong \frac{2\alpha r_e N}{\sigma_x} . \qquad (6)$$

This parameter serves as an approximate measure of backgrounds.

The luminosity can be rewritten in terms of only three free parameters: n_γ, σ_y, and the beam power, $P_B = N\gamma m c^2 f_c$,

$$L \cong \frac{1}{8\pi\alpha r_e m c^2} \frac{P_B n_\gamma}{\gamma \sigma_y} \qquad (7)$$

The diversity of approaches in Table III arises from different judgments about the following.

The ease of developing new and improving existing technology - SBLC and JLC-I(S) are the most conservative in this regard. They take advantage of over forty years of experience with S-band RF. NLC, JLC-I(C), and JLC-I(X) extend the basis of present day linacs, high peak power klystrons and modulators, to higher frequencies. Klystrons and accelerator structures must be developed for those frequencies. TESLA relies on substantial improvements in the cost and accelerating gradient of superconducting RF. VLEPP requires innovations to meet demanding tolerances and relies on novel beam dynamics in the linac and final focus. CLIC has stringent tolerances because of its high frequency, and the RF power source development by itself is a major undertaking comparable to the complete development of other colliders.

Costs - Cost reduction and cost control must be dominant considerations as designs are developed. New technologies promise significant, but uncertain, cost reductions. Older technologies have better established costs, but these tend to be high and must be lowered through engineering and mass production. The experience of the SSC, an accelerator based on mature technology and a detailed design, teaches us that present linear collider cost estimates should not be taken seriously.

Extension to higher energies - An $E_{CM} = 0.5$ TeV collider should be a step towards multi-TeV energies. High gradients and high RF frequencies tend to be better for reaching high energies with room temperature RF. NLC, JLC-I(X), and VLEPP are optimized for 0.5 - 1 TeV while it would be difficult to directly extend S-band colliders beyond 0.5 - 1 TeV. CLIC is a multi-TeV collider scaled down to 0.5 TeV for purposes of comparison. The energy reach of TESLA depends on how close the fundamental gradient limit of ~50 MV/m in Nb can be approached. This issue of extension to higher energies is discussed more in the next section.

Experimental backgrounds and center-of-mass energy spread - The effects of beamstrahlung have been captured in eq. (7) with a single parameter, n_γ. This parameter doesn't account for the energy spectra of photons, e^+e^- pairs, and hadronic events, and it doesn't account for the overlap of events in the detector. The complicated interface between collider and experiment cannot be reduced to a single number, and it is only through the ongoing studies of that interface that tolerable background levels can be estimated.

Tolerances and beam power - The trade-off is given in eq. (7). Increasing the beam power relaxes injection tolerances, beam position monitor precision, and pulse-to-pulse jitter in the final focus by allowing a larger σ_y. However, there are limits to beam power from efficiency and beam handling, collimation and accelerator protection.

Prototypes addressing beam dynamics and engineering will help narrow the range of choices. These prototypes include:
1.- A 500 MeV TESLA prototype to be constructed at DESY to demonstrate a gradient of 15 MV/m, to meet cost goals, and to test a high gradient superconducting linac with beam.
2.- A 450 MeV SBLC prototype that will test long pulse, high power, multiple bunch operation of an S-band linac.
3.- The Accelerator Test Facility at KEK that combines a 1.5 GeV, S-band linac with a prototype damping ring. The damping ring will produce beams with brightness, single bunch charge, and bunch train structure covering many of the colliders in Table III. New levels of tolerances, control of beam generated fields, extraction kicker stability, etc. will be reached in accomplishing this.
4.- Interaction region optics and stability will be studied at the Final Focus Test Beam at SLAC. In addition, strong field QED, the regime of beamstrahlung in high energy linear colliders, will be explored experimentally.
5.- A 540 MeV prototype NLC linac has the goals of constructing, reliably operating, and studying beam dynamics in an X-band linac.
6.- A ~500 MeV VLEPP prototype will test the klystrons, accelerator, and beam dynamics of that collider.
7.- A beam with the time structure of the CLIC drive beam will be generated by an RF gun, accelerated and used for demonstrating energy extraction at the CLIC Test Facility.

The lessons learned from continuing operation of the SLC and the answers to some broader questions will contribute as much as or more than these prototypes to determining the best approach for the next generation linear collider.

THREE ISSUES FOR THE FUTURE

There has been a change in the considerations that have dominated linear collider parameters. Early on the possibility of large accelerating gradients made high RF frequencies, laser-driven grating accelerators, and plasma accelerators attractive. Energy efficiency and beam dynamics have become more important now as the beam requirements and operating costs have become better understood. Complexity is a major factor that is still not being given the attention required. An assumption underlying the range of parameters in Table III is that each of colliders would be equally operable. There is no reason to believe this.

For example, the number of klystrons for the conventional, room temperature colliders in Table III (SBLC, JLC-I, and NLC), ranges from 1944 to 4180. Each klystron and its associated modulator is a major piece of equipment The factor of two range in the number of klystrons has the potential of making the difference between a collider that can be commissioned rapidly and will run reliably, and one that will not. Of course, more than just the number of klystrons is involved in assessing reliability; the properties of the components themselves are as important. The principles of reliability engineering and Quality Assurance need to be used to estimate the required reliability and provide a scientific basis for judging operability. This should become a part of the discussion soon.

Attention is focused on $E_{CM} = 0.5$ TeV for the next collider. Reaching much higher energies is one of the aspirations of the linear collider community, and one reason for supporting a 0.5 TeV collider is that it could serve as an intermediate prototype for a higher energy collider. Therefore, it is important to understand how the various approaches extend to higher energy. There are two different questions.

First, can a particular collider reach higher energies? There are ideas for extending the energy of all of the colliders in Table III up to $E_{CM} \approx 1$ TeV. Table IV gives parameters for three of them. The energy is increased by either: 1) keeping the length the same and increasing the peak power to the linac by a factor of four thereby doubling the gradient; or 2) doubling the length of the linac. At 0.5 TeV the trade-off between beam power and spot size in eq. (7) was exploited by TESLA and SBLC to increase the vertical spot size by increasing the beam power. That trade-off is not nearly as dramatic at 1.0 TeV. Everyone is relying on small spots to make luminosity.

Second, can an approach be extended to substantially higher energies, say $E_{CM} = 2 - 5$ TeV? It isn't productive to write down parameters of a collider two generations beyond the SLC; too much will be learned between now and then. Trends are clear, however; low beam power and small spots is the direction needed for multi-TeV colliders. The two different approaches to building a 0.5 TeV, superconducting RF and room temperature RF, have different outlooks for 2 - 5 TeV.

The advantage of superconducting RF at $E_{CM} = 0.5$ TeV is the ability to trade-off beam power and spot size, but that advantage has largely gone away by 1 TeV. Superconducting RF is a low gradient technology. There is a fundamental limit of about 50 MeV/m that comes from the breakdown of superconductivity when the surface field exceeds the critical field, and the practicalities of fabricating cavities makes the 25 MeV/m gradient of TESLA an ambitious but reasonable goal. The only way to extend the energy is to extend the length. A packing factor of 70% is hoped for, and the $E_{CM} = 1$ TeV accelerator is close to 55 km long for an active length of 40 km. The superconducting approach is not likely to take one far beyond $E_{CM} = 0.5$ TeV.

There is a factor of ten in the RF frequencies of the room temperature colliders in Table III, but there are many considerations in common including: multiple bunch energy control and higher mode damping and detuning needed for multiple bunches; precision beam position monitors, alignment, and steering algorithms needed for emittance preservation; optical corrections and jitter control for producing nanometer size spots; high reliability, low cost, efficient modulators, and economic fabrication of accelerator sections. Room temperature colliders are pushing technology and beam dynamics in the directions needed for multi-TeV collisions. In addition, the room temperature RF approach does not have a fundamental gradient limit forcing one to

Table IV Comparison of Linear Collider Parameters for
$E_{CM} = 0.5$ TeV and $E_{CM} = 1.0$ TeV

Parameter	TESLA		SBLC		NLC	
E_{CM} (TeV)	0.5	1.0	0.5	1.0	0.5	1.0
L (10^{33}cm^{-2}s^{-1})	7	10	4	6	8	20
Rep Rate (Hz)	10	5	50	50	180	120
Bunches/RF pulse	800	4180	125	50	90	67
N (10^{10})	5.15	0.91	2.9	2.9	0.65	1.3
$\gamma\epsilon_x/\gamma\epsilon_y$ (10^{-8}m)	2000/100	520/6.3	1000/50	1000/10	500/5	500/5
β_x^*/β_y^* (mm)	25/2	20/1	22/0.8	32/0.8	10/0.1	40/0.1
σ_{x0}/σ_{y0} (nm)	1000/64	325/8	670/28	572/9	300/3	425/2
σ_L (μm)	1000	500	500	500	100	100
Y	0.029	0.058	0.055	0.091	0.096	0.28
H_D	2.3	2.0	1.6	1.7	1.4	1.6
δ_B	0.03	0.03	0.03	0.07	0.03	0.08
n_γ	2.7	1.3	2.0	2.3	0.9	1.1
Load Grad.(MV/m)	25	25	17	34	38	76
Linac Length (km)	20	40	29.4	29.4	14	14
Number of Klystrons	1202	2404	2450	4900	1945	3890
Klystr Pk Pwr (MW)	3.25	3.25	150	150	94	188
Pulse Comp. Gain	--	--	--	2	4	4
P_B (MW)	16.5	15.2	7.3	5.8	4.2	8.0
AC Power (MW)	137	159	114	200	141	280
$2P_B/P_{AC}$	0.24	0.19	0.13	0.06	0.06	0.06

The $E_{CM} = 0.5$ TeV parameters are from G. Loew,[13] and the $E_{CM} = 1.0$ TeV parameters from B. Wiik.[26]

increase length to reach high energies. A room temperature, $E_{CM} = 0.5$ TeV collider has promise as an intermediate prototype for multi-TeV collisions.

There is world wide interest in building a large linear collider, and the interested parties have formed technical and scientific collaborations to that end. TESLA and the FFTB are two broad based, international collaborations pursuing linear collider development, and many of the other prototypes have participation outside of the home laboratory. The LC (Linear Collider) series of workshops has been a forum for accelerator physics discussions, and the workshops on Physics and Experiments at Linear Electron-Positron Colliders have played a similar role for the particle physics at these accelerators. A formal agreement on collaboration on accelerator development is being circulated, and the first collaboration council meeting should be in June, 1994.

A political mechanism for the support of a large linear collider is needed in addition to these scientist-to-scientist and laboratory-to-laboratory collaborations. Large fusion, space research, and high energy physics collaborations will set precedents for this. The former Director of the Office of Energy Research, William Happer, has commented on large scale high energy physics projects in the light of what is happening with ITER, the International Thermonuclear Experimental

Reactor.[27] Cost sharing and the difficulty of negating a treaty are important advantages of an international agreement for a large science project, but there are associated problems and costs. These include the difficulty of choosing a site, the time scale for formal, deliberate negotiations, and constituencies in the government that favor a slow time scale.

The biggest uncertainty we face is in the political arena.

ACKNOWLEDGEMENTS

Greg Loew has kindly let me reproduce his table of parameters from LC-93 (Table III). I have benefited from stimulating discussions about future linear colliders with Dave Burke, Pisin Chen, John Irwin, Bob Palmer, Tor Raubenheimer, and Ron Ruth. Finally, thanks to my SLC colleagues. It is an exciting accelerator to work on and a wonderful group of people to work with.

REFERENCES

1. LEP Design Report, CERN-LEP/84-01 (CERN,1984).
2. G. Bachy et al, Particle Accelerators 26, 19 (1990).
3. M. Tigner, Nuovo Cimento 37, 1228 (1965).
4. Table 9.2.4.1 in SLC Design Handbook (SLAC, Dec 1984).
5. C. Adolphsen et al, Proc of 1993 Part Accel Conf, SLAC-PUB-6255.
6. C. Adolphsen et al, Proc of 1993 Part Accel Conf, SLAC-PUB-6118.
7. N. J. Walker et al, Proc of 1993 Part Accel Conf, SLAC-PUB-6206.
8. K. Abe et al, Phys Rev Lett 70, 2515 (1993).
9. C. E. Adolphsen et al, 1991 IEEE Particle Accelerator Conference, 503 (1991).
10. P. Krejcik et al, Proc of 1993 Part Accel Conf, SLAC-PUB-6230.
11. This section is an update of R. H. Siemann, "Overview of Linear Collider Designs", Proc of 1993 Part Accel Conf, SLAC-PUB-6244.
12. ICFA Seminar held at DESY, May, 1993.
13. Greg Loew, private communication, Proc of LC93.
14. T. Raubenheimer, Proc of 1993 Part Accel Conf, SLAC-PUB-6117.
15. K. A. Thompson and R. D. Ruth, Proc of 1993 Part Accel Conf, SLAC-PUB-6154.
16. K. A. Thompson, C. Adolphsen and K. L. F. Bane, Proc of 1993 Part Accel Conf, SLAC-PUB-6153.
17. I. Wilson and W. Wuensch, Proc of the 1993 Part Accel Conf, CERN-SL/93-20.
18. P. Wilson, private communication.
19. Z. D. Farkas and P. B. Wilson, 1987 IEEE Particle Accelerator Conf, 1561 (1987).
20. P. B. Wilson, SLAC-PUB-3674 (1985).
21. P. B. Wilson, Z. D. Farkas and R. D. Ruth, SLAC AP-78 (1990).
22. V. Balakin, A. Novokhatski and V. Smirnov, Proc of 12th Int Conf on High Energy Accel, 119 (1983)
23. J. Irwin, private communication.
24. R. B. Palmer, Proc DPF Summer Study Snowmass '88, p. 613 (1988).
25. Pisin Chen, Photon-Photon Collisions, 418 (1992).
26. B. Wiik, presentation to LC93, Proc of LC93.
27. William Happer, AIP Conf Proc 272 (Proc XXVI Inter Conf on High Energy Physics) xxxvii (1993).

Manipulating Charged Particle Beams and Light by Means of Plasma

John M. Dawson

University of California, Los Angeles, Physics Department, 1-130 Knudsen Hall, Los Angeles, CA 90095-1547

Abstract. Plasmas can be used to focus, accelerate and bunch energetic electron beams. It is also possible to accelerate photons (shift their frequency), deflect them, focus them and bunch them in a plasma in much the same way as with electron beams. Some of the intriguing possibilities are presented here.

I have heard about Andy Sessler and his work for so long that I cannot remember when the first time was. However, it has only been in the last ten years or so that I have really gotten to know him through our common interests and work on plasma accelerators, plasma lenses and plasma light sources (the use of plasmas to upshift the frequency of light).

Because of our common interests, I spent a sabbatical at Lawrence Berkeley Laboratory with Andy's group in 1988-1989. I quite enjoyed that experience, except possibly for the problem of parking. One had to stack-park and leave a telephone number on the dashboard so that if your car was blocking someone, they could get hold of you to move it. At first, I was sure my car would be covered with dents by the time I left, but I was wrong; I do not think I got one dent there, but I do think I might qualify now as a parking attendant.

Considering the amount of traveling Andy and I did, it is surprising that we managed to collaborate at all. However, we did a surprising number of things together. We collaborated on plasma accelerators and lenses, we initiated a collaboration between Andy's group and Chan Joshi's at the University of California, Los Angeles to upshift the frequency of microwaves, and we even came up with a proposal for a method for doing DNA sequencing[1]. We even tried to get a proof-of-principal experiment started at Lawrence Livermore National Laboratory but we were unsuccessful at this. However, we did get a patent on the method. I think the method will work and I have since seen in the literature closely related techniques being successfully applied to other problems.

Today I am going to talk about ways to use plasmas to manipulate electron and light beams. This is still largely virgin territory and filled with many intriguing scientific problems. I believe it is an area with many potential practical applications.

© 1996 American Institute of Physics

Perhaps the first idea for manipulating electron beams by means of a plasma was the realization that plasma can act as a lens of unprecedented strength for focusing high energy electron and positron beams. In fact, when I was at LBL in 1988-1989, Andy came up with the idea of an adiabatic plasma lens[2].

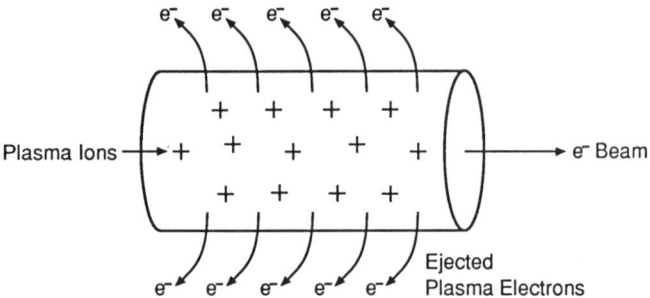

Figure 1. Plasma electrons being expelled from the region of a relativistic electron bunch neutralizing the electric field and leaving the bunch subjected to the magnetic pinching force which focuses the bunch.

The way such a plasma lens works is shown in Figure 1. For a relativistic bunch of electrons moving in vacuum, the electrostatic repulsive force and the magnetic pinching force cancel out to order γ^{-2}. If such a bunch moves into a plasma, the plasma electrons are repelled out by the electric force (the magnetic force essentially does not effect them because their velocity is very low). If the electron density of the beam is lower than the plasma electron density (underdense case), then the plasma electrons move out until the combination of beam electrons plus plasma electrons balances the ion density. The plasma is very effective at charge neutralization and shielding out electric fields. The plasma also tends to neutralize currents by generating return currents; this tends to shield the plasma from imposed magnetic fields; however, the magnetic shielding occurs over larger regions and is less effective than electric shielding. Thus, the beam sees no electrical repulsive force and is still subjected to a substantial magnetic pinching force which focuses the beam. If the electron density of the beam is higher than the plasma electron density (overdense case), then all the plasma electrons are repelled out of the beam region and the ion density provides an attractive electrical field that focuses the beam; there is no return current or magnetic shielding in the beam region in this case. In this latter case, the denser the plasma (so long as it is not as dense as the beam), the stronger the focusing force. In the adiabatic lens idea of Andy's, the plasma density rises as the beam is focused so that the beam density and the plasma density remain equal as the beam focuses. This gives the maximum focusing force at all times and as seen by the beam, it varies smoothly.

The plasma lens concept was first demonstrated 6-7 years ago at the Argonne National Laboratory by J. Rosenzweig[3]. Recently, an even more spectacular demonstration of it has been carried out at UCLA[4]. Figure 2 shows focusing by a factor of 6 in radius which was observed there.

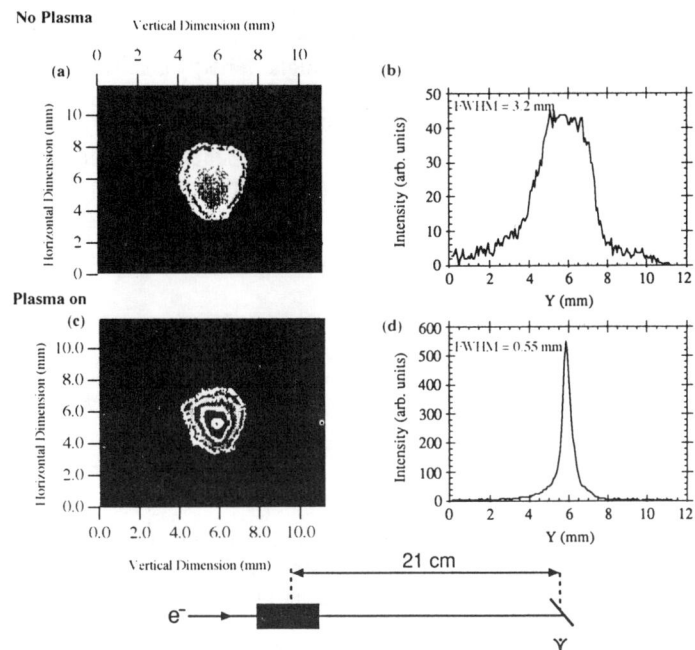

Figure 2. Results from the UCLA experiment showing the focusing of a plasma bunch by a plasma lens.

The second possibility I want to point out arises out of the recent demonstration of the Beat Wave Accelerator at UCLA[5], where electrons were accelerated from 2 to 30 Mev in a distance of about one cm. This experiment shows that coherent Langmuir waves with fields up to 3 Gev/M can be generated in a plasma at an electron density of 10^{16} per cm^3. The wavelength of the plasma wave is 300μ and they must be coherent over at least 30 wavelengths. The coherent waves probably last for between 10 to 100 picoseconds. The acceleration process will produce bunches of accelerated electrons. At the right distance these bunches will be 10s of microns long and have time durations of femtosecs; one bunch will come every picosec. In the present experiment, the bunches have a very low intensity because of the low current of the injected beam. They will be of low quality because the electrons will have a wide range of energies.

Another effect that influences bunching is the focusing properties of the waves. There are regions of the beatwave that are focusing and other regions that are defocusing. The electrons in the defocusing regions will be launched on diverging orbits by the wave, while those in the focusing regions will tend to concentrate on the axis of the beam. These effects will have to be included in any calculations of bunching; one may want to scrape off the diverging electrons.

If instead of injecting 2 Mev electrons, we had injected a beam at 300 Mev, the electrons riding the crests of the wave would have been accelerated to 330 Mev (γ = 660), and those at the trough would have decelerated to 270 Mev (γ = 540). After passing through the one cm wave section, the higher energy electrons overtake the lower energy electrons and bunching occurs. For the 300 Mev case, the maximum bunching occurs 45 cm beyond the accelerating region. At that point, and for quite a few cm around it, we will have very short intense bunches spaced every 300μ. The exact size of the bunches will depend on the quality of the beam, the quality of the plasma wave and the quality of the optics used; however, the size should be in the range of a few microns to a few tens of microns. The transit time of such a bunch, past a point, would be of the order of 10 femtosecs.

What can we do with such bunches? One of the simplest things to do is run them into a high Z target (say tungsten) and make femtosec pulses of X-rays from its K-shell excitation. We would then have a stroboscopic X-ray source consisting of femtosec pulses, with a pulse coming every picosec. You would have to find some way to separate the pulses after they pass through a system you are probing so that you can tell what time each pulse came; there are some possibilities of using plasma waves for this.

A second possibility is to backscatter a femtosec laser pulse off of the bunches. This could also give bursts of X-rays, one coming every half picosec. In this case, if one would ramp the energy of the bunches (later bunches have higher energy than earlier bunches) by passing them through a conventional accelerator cavity. This is done at an appropriate phase in its cycle just in front of the maximum bunching point. The X-rays from different bunches would have different frequencies and could be sent in different directions by reflecting them from a crystal diffraction grating.

The final topic to discuss is photon acceleration. Photons in a plasma behave like finite mass particles[6]. One can show that they behave like particles with a mass m = $h\nu/c^2$. They move with the group velocity of electromagnetic waves in a plasma ($v_g = c(1-\omega_p^2/\omega^2)^{1/2}$) and have momentum equal to their mass times γv_g. If the plasma has density gradients in it, then ω_p is a function of position, the photon mass varies with position, the group velocity and momentum are functions of position. One can show that a force acts on the photons due to the density gradients. One can treat the photons as particles, which are accelerated by these forces (one must include the mass changes of the photons as they move about in the plasma).

If the density variations are moving, then the reflection of waves can shift the frequency of the waves. For plasma waves with phase velocities close to the speed of light, these shifts can be quite significant. The plasma wave can exert forces

photons), or perpendicular to the motion of the photons (these forces cause deflection focusing and defocusing of the photons). Thus, photons can be manipulated by plasma waves in ways that are very similar to those for accelerating and focusing electrons.

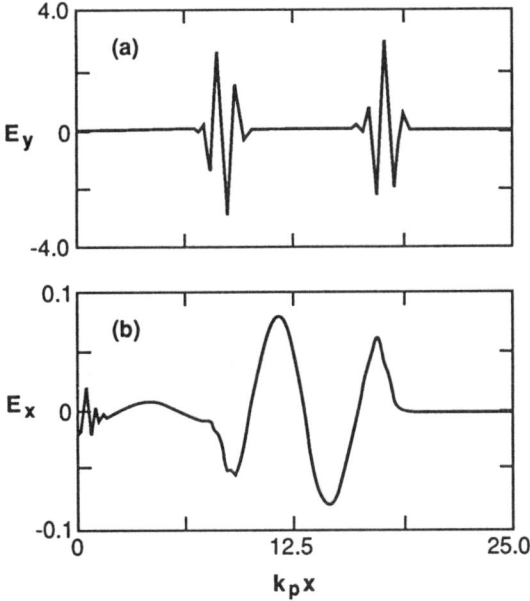

Figure 3. Top: Two short laser pulses propagating through a plasma separated by half a plasma wavelength. Bottom: The plasma wave wake emitted by the first pulse and absorbed by the second. The frequency of the first pulse decreases and that of the second pulse increases.

The effect is illustrated in Figure 3, where a short light pulse generates a plasma wave. A second electromagnetic pulse follows by half a plasma wavelength. The plasma wave it creates is out of phase with that from the first pulse and it cancels it out. The first pulse gives up energy to the plasma wave and its frequency shifts downward; the second pulse absorbs the plasma wave and its energy (frequency) increases. The plasma wave can also be generated by a bunch of electrons; in this way, we can transfer energy from electrons (which are easily accelerated) to photons and shift their frequency up[7].

The upshifting of the frequency of a light pulse by a plasma wave is illustrated in Figures 4a and 4b. Figure 4a shows a short light pulse which is being upshifted by a background plasma wave whose phase velocity matches the group velocity of the light. Figure 4b shows the frequency spectrum of the pulse at time t = 0 and at a later time, after its frequency has been shifted up. The narrow high intensity peek contains nearly all the energy and this peak shifts to higher frequencies in accordance with theory[6].

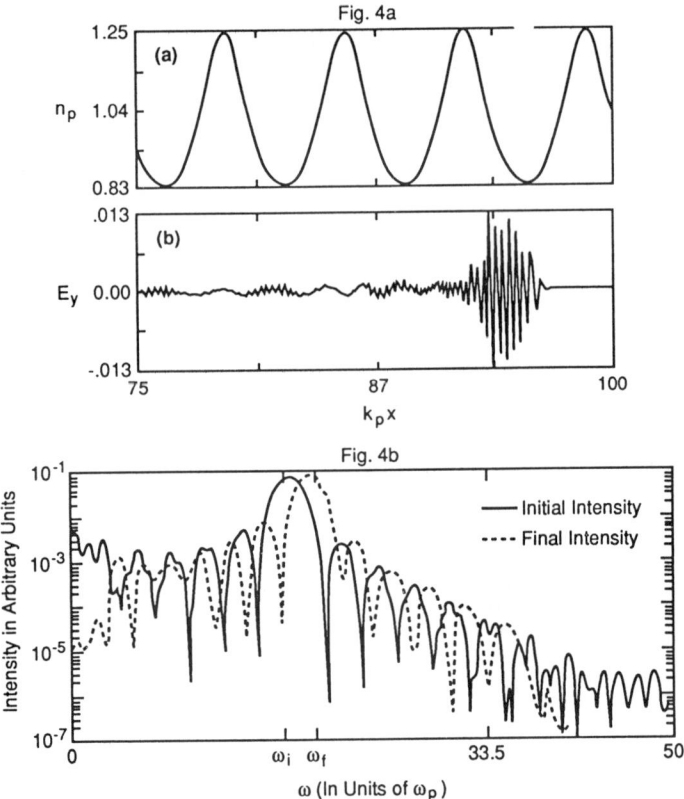

Figure 4. Top 4a: Electron density variation associated with a plasma wave; its phase velocity matches the group velocity of the light wave shown in the bottom of 4a. 4b: The initial and final intensity of the light pulse vs frequency.

References

1. Dawson, J. M., and Sessler, A., "DNA Base Pair Sequencer: Scanning Tunneling Microscope Plus Infrared Radiation," *Proc. of X-Ray Microimaging for the Life Sciences*, University of California, Berkeley, 1989, pp.108-113.

2. Sessler, A., *Phys. Fluids B* **2** 1326 (1990).
   Chen, P. K. Oide, A. M. Sessler, *Phys. Rev. Lett.* **4** 1231 (1990).

3. JRosenzweig, J. B., *Phys. Fluids B* **2** 1377, (1990).

4. Hairapetian, G., et al., *Phys. Rev. Lett.* **72** 2403 (1994).

5. Clayton, C. E., et al., *Phys. Rev. Lett.* **70** 37 (1993).

6. Wilks, S. C., Dawson, J. M., Mori, W. B., Katsouleas, T., and Jones, M. E., "A Photon Accelerator," *Phys. Rev. Lett.* **62** 2600 (1989).

7. Wilks, S. C., "Theory and Simulations of Novel Plasma-Based Accelerators and Light Sources," Ph.D. thesis, Physics Dept., University of California, Los Angeles (1989).

Excitation of Accelerating Wakefields in Inhomogeneous Plasmas

G. Shvets and J. S. Wurtele
Department of Physics and Plasma Fusion Center
Massachusetts Institute of Technology
Cambridge, MA 02139

April 4, 1995

Abstract

The excitation of the wakefields in an inhomogeneous plasma by a short laser pulse is investigated theoretically. A general equation for the wake excitation in transversely non-uniform plasma is derived. This equation is applied to the step-function density profile model of hollow channel laser wakefield accelerator [1]. A more realistic model, in which the transition between the evacuated channel and the homogeneous surrounding plasma occurs over a finite radial extent, is then analyzed. It is shown that the excited channel mode can interact resonantly with the plasma electrons inside the channel wall, leading to secular growth of the electric field. This eventually results in wave-breaking and the dissipation of the accelerating mode. We introduce an effective quality factor Q for the hollow channel laser wakefield geometry. This resonance limits the number of electron bunches that can be accelerated in the wake of single laser pulse.

PACS numbers: 52.40.Nk, 52.35.Mw, 52.75.Di

1 Introduction

Since the 1950's Andy Sessler has made continuing significant contributions to accelerator physics. Innovation and the development realizable concepts are among the hallmarks of his scientific research. An informal atmosphere and a broad interest in science, culture and human rights characterize the intellectual environment of his group. His style of doing physics has influenced not only his students, but, through them, their students as well. It is thus a great pleasure for us to dedicate this paper to Andy Sessler.

One of the applications of the short-pulse laser-plasma interactions that has drawn considerable theoretical [2, 3, 4] and experimental [5] attention is to use the plasma as an accelerating medium. The idea behind the laser-plasma accelerator is very simple: a short intense laser (laser wakefield accelerator) or a superposition of two lasers pulses with slightly different frequencies (laser beat-wave accelerator) can be used to excite a longitudinal plasma wave that can accelerate an electron bunch to high energies. The ability of the plasma to support accelerating gradients as high as a few Gev/m has been already demonstrated experimentally [5, 6] and even higher gradients are theoretically achievable. At the same time, a number of physical phenomena limit the energy of the accelerated electron bunch. Diffraction and depletion reduce the distance that the laser pulse can travel through the plasma while maintaining the large amplitude of the electromagnetic field required to excite the wake, and dephasing [7] limits the distance over which the electron bunch remains at the accelerating phase of the plasma electric field.

Several schemes have been proposed to overcome diffraction. Relativistic guiding [8, 9, 10, 11] relies on the energy dependence of the plasma frequency, ω_{p0}^2/γ, where $\omega_{p0}^2 = 4\pi n_0 e^2/m$, with n_0 the unperturbed palsma density, $-e$ the electron charge, and m the electron mass. The relaticistic factor $\gamma = \sqrt{1 + \vec{p}\cdot\vec{p}/m^2c^2}$, with c the speed of light. The electron momentum $|\vec{p}|$ will be largest where the laser pulse is most intense, and therefore the plasma frequency will be lower there, and the pulse will generate a nonlinear index of refraction which is larger at the center of the pulse than at the pulse edges. Analysis has shown [9] that, in steady-state, relativistic guiding can focus the a long laser pulse with frequency ω whenever the total power is greater than $P_c = 16.2(\omega/\omega_{p0})^2$ GW.

For the short pulses (of order a plasma wavelength) envisioned in many wakefield accelerators, however, relativistic guiding is substantially reduced [7]. This is due to the tendency of the ponderomotive force from the front of the pulse to push plasma electrons forward and generate a density increase which balances the relativistic mass increase. The plasma frequency then has no transverse variation and cannot optically guide the laser pulse. Relativistically guided long laser pulses suffer from Raman forward and sidescatter instabilities [12, 13]. This instability leads to the break-up of the pulse into small pulselets of order one plasma wavelength. The utility of using these pulselets themselves for acceleration is at present unclear [14].

An alternative scheme which has been investigated theoretically [1, 15] and experimentally [6] envisions guiding the laser pulse with a plasma density channel. The channel should have a higher density on the outside than on the inside, giving it an index of refraction which decreases from the channel axis. A fixed plasma channel is analogous to an optical fiber, and its guiding properties can be similarly analyzed [6]. The plasma channel can be used to guide short pulses for particle acceleration, and has been theoretically studied for parabolic density variation [15]-[17] and for hollow channels [1].

Analytic progress was obtained in many studies [15]-[17] by restricting the investigation to the case that the plasma is almost homogeneous over distances where the ponderomotive pressure of the laser is substantial (roughly equal to the laser spot size). In the approximation the plasma inhomogeneity can be neglected when calculating the density perturbation of the plasma induced by the laser (i.e., the wake). The plasma inhomogeneity was included deriving the unperturbed wave equation for the laser. The analytically tractable form of the unperturbed laser eigenmodes in the parabolic channel was the primary motivation for assuming this particular transverse density profile in Ref. [16]. The detailed knowledge of the transverse eigenmodes facilitated the derivation of a coupled-mode differential equation in [16] which described the spatio-temporal evolution of an arbitrary transverse distortion of the laser pulse in the limit of $P/P_c \ll 1$.

The goal of this paper is to develop a general formalism for analyzing wakefield generation in the plasma without apriori assumptions on the ratio of the plasma and laser transverse gradients. The motivation (as well as the starting point for most of the calculations presented here) can be found in a previous study of the hollow channel wakefield accelerator (HCLWA) [1]. In this scheme an evacuated channel in the plasma serves as an optical fiber which guides the laser pulse over many Rayleigh lengths. At the same time, the ponderomotive force of the laser excites wakefields at the surface of the channel, which extend to the center where they can be used for particle acceleration. As was first pointed out in Ref. [1], the accelerating mode of the HCLWA is transversely uniform. This property of the wake is very attractive for particle acceleration since transverse inhomogeneities of the accelerating gradient introduce unwanted energy spreads and impose stringent limitations on the transverse emittance. Hence, the hollow plasma channel *decouples* the transverse profile of the laser pulse (which is in general non-uniform and can not be easily manipulated) and the transverse profile of the accelerating mode. In addition, the hollow channel provides the much needed focusing to the laser pulse allowing it to propagate over many Rayleigh lengths.

The previous analysis and numerical study of the HCLWA [1] was mostly restricted to the case where the interface between the evacuated channel and the uniform plasma is infinitely sharp, although a general equation for the excitation of the wake in the inhomogeneous plasma in slab geometry was derived in the Appendix of Ref. [1]. In this paper we explore the implications of this equation to the more realistic case of a finite thickness interface (which we refer to as the finite thickness channel wall), studied for the first time in Ref.[18]. We find that the accelerating mode of the channel, ponderomotively excited by the laser pulse, resonates with the plasma electrons inside the channel wall. We analyze the temporal behavior of the electric fields inside the channel wall and find that (within the constraints of the linear theory) the transverse electric field at the resonant location grows secularly with time even after the laser pulse has gone by. This growth subsequently leads to the wavebreaking and dissipation of the wake. We introduce an effective quality factor Q of the plasma channel and find limits on the number of the electron bunches that can be injected into the wake created by a single laser pulse.

As was mentioned in Ref. [1], the idea of the hollow channel accelerator bears some resemblance to the plasma fiber accelerator proposed by Tajima. An important difference between the two approaches is that in the case of the plasma fiber accelerator plasma was assumed overdense while in the case of the HCLWA the plasma is underdense. One of the serious shortcomings of the plasma fiber accelerator was the resonant absorption of the laser in the plasma. In the case of the HCLWA, nowhere in the plasma does the frequency of the propagating laser match the local plasma frequency. However, a more subtle effect takes place: the excited wake has the frequency *lower* than the plasma frequency of the surrounding plasma. This leads to the resonant absorption of the wake. While this absorption of the *wake* is less deleterious that the resonant absorption of the *laser* (since most of the dissipation occurs behind the pulse), it should be certainly taken into account when designing an accelerator.

This paper is organized as follows: Section 2 introduces the notation and reviews the basic equations governing the excitation of a plasma wake in an inhomogeneous plasma. We also review relevant results, previously obtained in Ref. [1], for the wake excitation in the hollow channel with the step-function dependence of the plasma density on the transverse coordinate. This provides the theoretical background for our study of wake excitation in the channel with finite thickness walls. Section 3 analyzes the eigenmodes of a hollow channel with walls of thickness $d = a\delta$, where a is the width of the hollow channel and $\delta \ll 1$. For the case of a linear density variation in the walls, we calculate to the lowest order in δ the eigenfrequency of the accelerating channel mode. We find that the eigenmode of a hollow channel with finite thickness walls is singular at the resonant location where the local plasma frequency matches the eigenfrequency of the channel. One of the consequences of the singular character of the mode is that it dissipates over time. The rate of the mode dissipation is calculated to te lowest order in δ in Section 4. In reality, the singular mode would take an infinite period of time to develop and would be arrested by various non-linear phenomena such as wavebreaking. This motivates the calculation of the temporal evolution of the fields inside the channel wall, in the framework of the linear theory, which is carried out in Section 5. Section 6 summarizes our results and outlines directions for the future work.

2   Basic Formalism and the HCLWA

A number of simplifying assumptions make the problem of wakefield generation in the inhomogeneous plasmas amenable to analysis. We assume that the laser pulse is much shorter than the responce time of the ions, thus reducing the problem to the interaction of the intense laser with the plasma electrons in the neutralizing background of the immobile ions. We further assume that the plasma is homogeneous in $z-$ direction—the direction of the laser propagation—and restrict ourselves to slab geometry. The unperturbed local plasma frequency defined as

$$\omega_{p0}^2(x) = \frac{4\pi e^2 n_0(x)}{m}, \tag{1}$$

where $n_0(x)$, the unperturbed plasma density, varies only in the $x-$direction. The plasma is taken to be underdense, so that $\omega_p^2 \ll \omega_0^2$, where ω_0 is the laser frequency. This allows us to separate time scales into the fast scale of order a laser period, $1/\omega_0$, and slow scale of order a plasma period, $1/\omega_{p0}$. Plasma electrons move in the combined (fast scale) electric and magnetic fields of the laser and the (slow-scale) fields induced by the laser in the plasma. Since the plasma frequency is assumed to be much smaller than that of the laser, we can time-average over the laser period so that electrons are only driven by the poderomotive force of the laser.

An important point to note is that, for the purpose of designing the best wakefield accelerator, the highest laser field amplitude $|a| \approx e|A|/mc^2$ (see Eq. 3), where A is the vector potential of the laser field, *is not* desirable. For the best acceleration one should avoid over-bunching the plasma, so as to prevent wave-breaking, profile steepening and related nonlinear effects that are undesirable for particle acceleration. Increasing a also inhances pump depletion [18], yet a very small a would inefficiently excite a weak wake. We will assume that

$$a \leq 0.3. \tag{2}$$

and use the weakly relativistic approximation [12]. Consistent with this approximation, the calculation of the density perturbation and relativistic index of refraction will be performed to second order in the laser field, neglecting the terms proportional to a^4.

The laser field is described by a vector potential $\vec{A}$, with a varying on a slow scale:

$$\vec{A} = \frac{mc^2}{2e}\vec{a}(\vec{x}_\perp, z, t)e^{i(k_0 z - \omega_0 t)} + c.c., \qquad (3)$$

where

$$\vec{a} \cdot \vec{e}_z = 0. \qquad (4)$$

The spot size of the laser is assumed to be much larger than the wavelength λ_0, validating the assumption of transverse polarization in Eq.(4). Under the weakly relativistic assumption (2), the peak particle displacement in the transverse plane is negligible compared with the spot size of the laser.

Separating electron velocity into fast and slow parts, using the conservation of transverse canonical momentum, and averaging over the fast timescale yields an equation of motion for the slow-scale plasma velocity $\vec{v}$:

$$\frac{\partial \vec{v}}{\partial t} = \frac{e}{m}(\vec{E} + \nabla f), \qquad (5)$$

where f is the ponderomotive potential given by

$$f = -\frac{mc^2}{4e}|\vec{a}|^2, \qquad (6)$$

and E is the electric field induced in the plasma. In deriving Eq.(5) we have used Eq.(2) to neglect the nonlinear terms of order a^4.

As is known from the earlier work on interaction of obliquely incident p-polarized lasers with inhomogeneous plasmas [19, 20] that for stratified inhomogeneous plasma calculations are more easily performed in terms of the magnetic field $\vec{B}$. Fourier transforming the Faraday's law in time, and introducing a new quantity

$$\vec{P} = \frac{ic}{\omega}\vec{B}$$

results in

$$\vec{E} + \frac{4\pi i}{\omega}\vec{j} = \nabla \times \vec{P}. \qquad (7)$$

Combining Eq.(7) with Eq.(5) and recalling that the current in the electron plasma is given by $\vec{j} = en_0\vec{v}$ yields:

$$\vec{E} = \frac{\omega_{p0}^2(x)/\omega^2}{1 - \omega_{p0}^2(x)/\omega^2}\nabla f + \frac{\nabla \times \vec{P}}{1 - \omega_{p0}^2(x)/\omega^2}. \qquad (8)$$

Later in the section we show that for the case of a homogeneous plasma vector $\vec{P}$ goes to zero and the wake becomes electrostatic.

¿From Maxwell's Equations,

$$\nabla \times \nabla \times \vec{E} + \frac{1}{c^2}\frac{\partial^2 \vec{E}}{\partial t^2} = -\frac{4\pi}{c^2}\frac{\partial \vec{j}}{\partial t}. \qquad (9)$$

Combining Eq.(9) with Eq.(8) we obtain:

$$\nabla \times \nabla \times \left(\frac{\nabla \times \vec{P}}{1 - \omega_{p0}^2(x)/\omega^2} + \vec{\nabla} f \frac{\omega_{p0}^2(x)}{\omega^2} \frac{1}{1 - \omega_{p0}^2(x)/\omega^2} \right) =$$

$$= \frac{\omega^2}{c^2}\nabla \times \vec{P}. \qquad (10)$$

Equation (10) can be simplified to a scalar differential equation in instances where the plasma has planar or azimuthal symmetry. In these limits, the outer curl in Eq.(10) can be removed and Eq.(10) is substantially simplified:

Case 1. Slab geometry, $n_0 = n_0(x)$, $f = f(x, z)$ and the TM space-charge mode is exited, with $\vec{P} = \vec{e}_y P(x, z)$.

Case 2. Cylindrical geometry, $n_0 = n_0(r)$, $f = f(r, z)$. Also, the TM wave is generated with $\vec{P} = \vec{e}_\phi P(r, z)$.

The physical reason for this simplification is that, from the symmetry of the cylindrical and slab geometries, magnetic field is unidirectional. In this paper we concentrate on case 1. The extension to azimuthally symmetry is straightforward.

By introducing $\vec{P} = \vec{e}_y P(x, z)$, Eq.(10) is reduced to

$$-\nabla^2 P + \frac{\partial P}{\partial x}\frac{\partial}{\partial x}\ln\left(1 - \frac{\omega_{p0}^2(x)}{\omega^2}\right) - \frac{\omega^2 - \omega_{p0}^2(x)}{c^2}P =$$

$$= -\frac{\partial f}{\partial z}\frac{\partial}{\partial x}\ln\left(1 - \frac{\omega_{p0}^2(x)}{\omega^2}\right). \qquad (11)$$

Equation (11), in a slightly modified form, was used for analyzing the resonant absorption of obliquely incident p polarized radiation by inhomogeneous plasma [19, 20, 21]. It is used for the first time, to our knowledge, for studying pondcromotivcly induced plasma wakes for particle acceleration.

We will concentrate on the wakes left by nonevolving laser pulses moving with speeds close to the speed of light. This is equivalent to considering all the quantities of interest as functions of x and a single longitudinal variable $\zeta = ct - z$. Then, since

$$\frac{\partial}{\partial z} = -\frac{\partial}{c\partial t}, \qquad (12)$$

Eq.(11) simplifies to

$$-\frac{\partial^2 P}{\partial x^2} + \frac{\partial P}{\partial x}\frac{\partial}{\partial x}\ln\left(1 - \frac{\omega_{p0}^2(x)}{\omega^2}\right) + \frac{\omega_{p0}^2(x)}{c^2}P =$$
$$= -i\frac{\omega}{c}\tilde{f}\frac{\partial}{\partial x}\ln\left(1 - \frac{\omega_{p0}^2(x)}{\omega^2}\right), \qquad (13)$$

where

$$\tilde{f}(x,\omega) = \int_{-\infty}^{+\infty} dt e^{i\omega t} f(x,t). \qquad (14)$$

Eq.(13) holds for any transverse density profile $\omega_{p0}^2(x)$, including those with density discontinuities. Equation (8) and a knowledge of the $\vec{P}(x,\zeta)$ is sufficient for computing the electric field.

For continuous density profiles Eq.(13) can be solved for P using the following boundary condition:

$$\lim_{x \to \pm\infty} P(x) = 0. \qquad (15)$$

In a homogeneous plasma ($\frac{\partial}{\partial x}\omega_{p0}^2(x) = 0$) it is evident that $P = 0$ is the only solution satisfying Eqs.(13-15). This implies that the plasma wake is electrostatic in a homogeneous plasma.

For discontinuous density profiles (e.g., a hollow channel or one with density steps) one has to match, at the discontinuity, the solutions from adjacent regions of continuous density. The usual way of matching the solutions at discontinuities is to find invariants that remain continuous across the density jump. Close inspection of Eq.(13) gives the following continuity conditions:

$$P(x) \to \text{continuous} \qquad (16)$$

$$\frac{\frac{\partial P}{\partial x} + i\frac{\omega}{c}\tilde{f}}{1 - \frac{\omega_{p0}^2(x)}{\omega^2}} \to \text{continuous}, \qquad (17)$$

where one can show that Eq.(17) is equivalent to demanding the continuity of E_z and Eq.(16) is equivalent to balancing the jump in E_x with the surface charge. In deriving the continuity conditions (16-17) we have assumed that ponderomotive potential f is continuous across the boundary. Since f is proportional to the amplitude of laser radiation (which is guided by the channel itself), it is clear that some discontinuity of f at the boundary is unavoidable if, for example, the laser is polarized in x direction. Yet, since $\omega_p^2 \ll \omega_0^2$, this discontinuity can be neglected compared to the discontinuity of $\frac{\partial P}{\partial x}$. Below we apply the general formalism developed here to a hollow channel.

The eigenmodes of an arbitrary channel of transversely inhomogeneous plasma can be found by solving Eq.(13) with $f = 0$, which then becomes an eigenvalue equation for wakes, moving with the speed of light, varying as $e^{-i\frac{\omega}{c}(ct-z)}$, where ω is an eigenvalue of the RHS of Eq.(13). For an arbitrary density profile $n_0(x)$ Eq.(13) is very hard to study analytically (as well as numerically), as will be explained in Section 3.

Analytical progress can be made for a simple hollow channel geometry where

$$\begin{aligned} \omega_{p0}^2(x) &= 0 \quad \text{for} \quad |x| < a \\ \omega_{p0}^2(x) &= \omega_p^2 \quad \text{for} \quad |x| > a \end{aligned} \quad (18)$$

We now look for the eigenfunctions $P(x)$, satisfying Eq.(13) and continuity conditions (16-17), with $f = 0$, at $x = \pm a$. The symmetry of the problem with respect to an $x \to -x$ transformation assures that the eigensolutions will be either even or odd in x. In fact, in the rest of this chapter we will be only considering the *odd* modes since they correspond to *even* accelerating gradients and are excited by laser pulses which are symmetric in x. The *even* modes that are excited by an initially offset laser pulse or one that has degraded through instabilities (such as are examined in Ref. [16]), can be analyzed by techniques similar to those given here.

The only eigenmode of the hollow channel is given by

$$\begin{aligned} P &= A\frac{x}{a} \quad \text{for} \quad |x| < a \\ P &= Ae^{-k_p(x-a)} \quad \text{for} \quad x > a \\ P &= -Ae^{k_p(x+a)} \quad \text{for} \quad x < -a, \end{aligned} \quad (19)$$

where $k_p = \omega_p/c$ and A is an arbitrary constant. The eigenfrequency of this mode, found by applying the continuity condition (17), is given by

$$\omega_{ch} = \frac{\omega_p}{\sqrt{1 + k_p a}}. \quad (20)$$

Using Eq.(8) we find the accelerating gradient inside the channel to be transversely uniform:

$$E_z = \frac{A}{a}. \quad (21)$$

This attractive property of the hollow channel laser wakefield accelerator was first realized in Ref. [1]. Transverse inhomogeneities in the accelerating field introduce unwanted energy spread in the bunch and impose stringent limitations on the transverse emittance. An analysis that did not approximate $v_g = c$ would find gradients of order ω_p^2/ω_0^2.

It is easy to show that the density perturbation associated with the eigenmode (19) is equal to zero inside the plasma and creates a surface charge layer at the edge of the plasma. The fields inside the channel are the fringing fields from this surface charge. Plasma particle trajectories are ellipses in $x-z$ plane, with an aspect ratio equal to $k_p a$.

To relate the amplitude of the wake A to the strength of the laser driver f, the continuity condition (17) is applied to the eigensolution (19), resulting in

$$\frac{A}{a} = \omega_p^2 \frac{i\omega \bar{f}}{c(1+k_p a)(\omega^2 - \omega_{ch}^2)}, \qquad (22)$$

where

$$\bar{f} = \tilde{f}(x = a). \qquad (23)$$

Assuming that $f = f(\zeta, x)$, the accelerating field inside the channel can be expressed as

$$E_z(\zeta) = \frac{k_p^2}{(1+k_p a)} \int_{-\infty}^{\zeta} d\zeta' \frac{\sin k_{ch}(\zeta - \zeta')}{k_{ch}} \frac{df(\zeta', x = a)}{d\zeta'}. \qquad (24)$$

Note that this relation between the ponderomotive driver and the wake amplitude is identical to that in a homogeneous plasma with the small difference that k_{ch} is replaced by k_p in the sine and the coupling constant factor $(1+k_p a)$ is replaced by unity. Of course, for the homogeneous plasma the ponderomotive force is evaluated on axis (i.e., at its peak), while for the channel it is evaluated at the channel wall (corresponding to a value less than the peak). Another, somewhat unexpected consequence of Eq.(24), is that the amplitude of the wake is driven by the *longitudinal* rather than transverse shape of the ponderomotive potential f. Hence, despite the significant ponderomotive pressure a narrow laser pulse exerts on the plasma *transversely*, it still has to be short to create substantial wakes inside the channel. As Eq.(24) indicates, the accelerating gradient inside the hollow channel is transversely uniform and determined by the amplitude of the laser pulse at the edge of the channel. Thus, the plasma channel effectively decouples the transverse profile of the laser from the transverse profile of the wake.

3 Thin-Wall Hollow Channel

In Section 2 we were able to calculate exactly the plasma wake driven by the ponderomotive force in a hollow channel. This hollow channel with a step function density profile is clearly an idealization. In a realistic hollow channel plasma density will rise continuously from zero (inside the channel) to a constant value far away from the channel. To quantitatively anticipate the new physics exhibited by a smooth channel, imagine that the thickness of the "channel wall" (the distance over which plasma density changes from 0 to n_0) is much smaller than the size of the channel. Then, we expect (and will show later in this paper) that the eigenfrequency of the channel mode, ω_{ch} does not change much from that given in Eq. (20).

Since
$$0 < \omega_{ch} < \omega_p, \qquad (25)$$
there exists a location x_r inside the channel wall where the channel frequency matches the local plasma frequency:
$$\omega_p(x_r) = \omega_{ch}. \qquad (26)$$
A resonant enhancement of the electric field is expected at this location. The coefficient multiplying the first derivative of P in the eigenmode equation (13) becomes singular.

Expanding the coefficients of Eq.(13) in the vicinity of $x = x_r$, to the lowest order in $(x - x_r)$, and defining L by
$$\frac{d\left(\frac{\omega_p^2}{\omega_{ch}^2}\right)}{dx}(x = x_r) = \frac{1}{L}, \qquad (27)$$
gives, (with $f = 0$),
$$-P'' + \frac{P'}{x - x_r} + \frac{\omega_{ch}^2}{c^2} P \left(1 + \frac{x - x_r}{L}\right) = 0, \qquad (28)$$
where a prime denotes $\frac{d}{dx}$.

A general solution of Eq.(28) in the vicinity of $x = x_r$ can be expressed as a series in $\xi = (x - x_r)$:
$$\begin{aligned} P &= A\left(\frac{k_{ch}^2 \xi^2}{2} \ln(k_{ch}\xi) + 1 + \cdots\right) \\ &+ B(k_{ch}^2 \xi^2 + \cdots), \end{aligned} \qquad (29)$$
where only the first terms in the power series are given, and A and B are numerical constants. The solution (29) inside the ramp region should match at the ramp boundaries with the corresponding solutions inside the channel and inside the homogeneous plasma (given by Eq.(19)).

A simple illustration of how the matching occurs is helpful: we assume, for simplicity, that $k_p a = 1$. Then, using Eq.(20), we find that $k_{ch} = k_p/\sqrt{2}$, and the boundaries of the ramp are located at $x_b = \pm L$. We further assume that

$L \ll a$. Then, to guarantee *amplitude* matching with solutions of Eq. (19), the constant A in Eq.(29) should be equal to the A in Eq.(19). On the other hand, the constant B has to be chosen so as to match the *derivatives* of the solutions of (29) with those of (19). It is a matter of simple algebra to see that $B = a/L$. Notice that $B \to \infty$ as the wall thickness L approaches zero, yet B *does not* contribute to the amplitude of P in this limit! The increase of B in the limit of an infinitely thin wall is related to the fact that there is a jump in transverse field E_x due to the surface charge at the edge of the plasma. As a result, for a step function density profile P has a discontinuous derivative at the channel edge (as seen from Eq.(19)), while for a channel with a finite-thickness wall P has a smooth maximum at $x = x_r$.

It is evident from Eq.(8) that the axial and transverse electric fields diverge as $x \to x_r$:

$$E_z \propto k_{ch} A \ln(k_{ch}\xi)(k_{ch}L) \qquad (30)$$
$$E_x \propto k_{ch} A (k_{ch}\xi)^{-1}(k_{ch}L). \qquad (31)$$

The merit of using P (or B_y) to characterize the eigenmodes of an inhomogeneous plasma, instead of the electric field, is now very clear. Both components of $\vec{E}$ are divergent at the resonant location, which would make an equation for E, equivalent to Eq.(13), hard to analyze even numerically. On the other hand, P stays finite and differentiable at $x = x_r$, as seen from the assymptotic expansion (29). The robustness of P to small changes in the wall thickness suggests that the eigenfrequency of the channel ω_{ch} does not change much from (20) if the wall thickness is small.

We see that, as the channel wall becomes thinner ($L \to 0$), for a fixed A (which corresponds to the accelerating gradient inside the channel, as in Eq.(21)), the amplitude of the singularity decreases, vanishing for a step function density profile. Obviously, the electric field at $x = x_r$ neither builds up to high values instantaneously, nor, because of nonlinear effects will it ever become infinite. To understand the dynamics of the build up of the electric field, we examine in Section 5 the time-dependent exitation the wake.

But before developing the time-dependent theory, we perturbatively examine the eigenmodes of a hollow channel with thin walls. This analysis is needed to: (a) develop an operator approach to the problem of wake excitation in an inhomogeneous plasma, and (b) to perturbatively calculate the structure of the eigenmode inside the channel wall, proving that neither the eigenfrequency, nor the accelerating mode itself, are greatly affected by the finite thickness of the channel wall. Numerical and analytical solutions to the eigenvalue equation given by the LHS of Eq.(13) are obtained below for a plasma with a linearly tapered density (from 0 to n_0) over a distance short compared with the size of the channel, a.

By defining a linear operator

$$\mathcal{L} = -\epsilon(x,\omega)\frac{\partial}{\partial x}\left(\frac{1}{\epsilon(x,\omega)}\frac{\partial}{\partial x}\right) + \frac{\omega_p^2(x)}{c^2}, \tag{32}$$

where

$$\epsilon(x,\omega) = 1 - \frac{\omega_p^2(x)}{\omega^2}, \tag{33}$$

the LHS of Eq.(13) can be symbolically rewritten as

$$\mathcal{L}P(x) = 0, \tag{34}$$

where the frequency dependence of $\mathcal{L}$ is suppressed for notational convenience. It is straightforward to prove a useful identity which holds for arbitrary odd functions $\psi_1(x)$ and $\psi_2(x)$:

$$\int_0^{+\infty} dx \frac{\psi_1(x)}{\epsilon(x,\omega)}\mathcal{L}\psi_2(x) = \int_0^{+\infty} dx \frac{\psi_2(x)}{\epsilon(x,\omega)}\mathcal{L}\psi_1(x). \tag{35}$$

Thus, with the weighting factor $\frac{1}{\epsilon(x,\omega)}$, $\mathcal{L}$ is a Hermitian operator (assuming that ω is real). Therefore, it has a complete set of real eigenvalues λ_n and corresponding eigenfunctions ψ_n (that can be also chosen to be real):

$$\mathcal{L}\psi_n(x) = \lambda_n \psi_n(x), \tag{36}$$

where $\lambda_n = \lambda_n(\omega)$ since the operator $\mathcal{L}$ is frequency dependent. Note that the lower limit of integration is chosen as 0 (instead of $-\infty$) since, as was mentioned earlier, only odd modes are considered in this paper.

The orthogonality condition for the eigenfunctions is derived from Eqs.(35) and (36):

$$\int_0^{+\infty} \frac{dx}{\epsilon(x,\omega)}\psi_{n_1}\psi_{n_2}(x) = 0 \quad \text{for } n_1 \neq n_2$$
$$= U_{n_1} \quad \text{for } n_1 = n_2, \tag{37}$$

where U_n is the weighting coefficient. Note that all the integrals are defined in a principle value sense since $\epsilon(x,\omega)$ can contain poles and that the index n *does not* imply a purely discrete spectrum. In fact, the spectrum contains both discrete ($\lambda < k_p^2$) and continuous ($\lambda > k_p^2$) eigenvalues. For continuous part of the spectrum the eigenfunctions, obviously, are not normalizable.

The mode we are presently concerned with is the self-supported wake, corresponding to $\lambda = 0$, as can be seen from Eq.(34). The unperturbed solution for a step-function profile ψ_0 is given by Eq.(19), and its eigenfrequency

$\omega_u = \omega_{ch}$. We now assume that the thin-wall dielectric function is given by

$$\epsilon = 1 \quad \text{for} \quad |x| < a(1 - \delta/2)$$
$$\epsilon = 1 - \frac{\omega_p^2(x - a + a\delta/2)}{\omega^2 a \delta} \quad \text{for} \quad a(1 - \delta/2) < |x| < a(1 + \delta/2)$$
$$\epsilon = 1 - \frac{\omega_p^2}{\omega^2} \quad \text{for} \quad |x| > a(1 + \delta/2), \tag{38}$$

where $\delta \ll 1$ is the ratio of a channel thickness to the half-width of the channel. Note that the *average* position of the channel wall remains unchanged, at $x = a$, to exclude the obvious effect of the channel width on the eigenfrequency. With these definitions it is straightforward to estimate, to linear order in δ, the influence of density ramp on the wake eigenmode.

Equation (34) for the wake eigenfunction ψ can be rewritten as

$$\frac{\partial}{\partial x}\left(\frac{1}{\epsilon(x,\omega)}\frac{\partial \psi}{\partial x}\right) = \frac{\omega_p^2(x)}{c^2 \epsilon(x,\omega)}\psi. \tag{39}$$

The solutions ψ in the regions $0 < x < a(1 - \delta/2)$ and $x > a(1 + \delta/2)$ are known and given by Eq.(19). Therefore, Eq.(39) can be integrated accross the density ramp to yield a jump in $\psi'/\epsilon \propto E_z$. Note that in the case of a step function density profile the size of the density ramp is equal to zero, and, thus, the jump also vanishes, as is required by Eq.(17).

The function ψ is found by writing it as

$$\psi = \frac{A_1 x}{a(1 - \delta/2)} \quad \text{for} \quad x < a(1 - \delta/2)$$
$$\psi = 1 - B(x - x_r)^2 \quad \text{for} \quad a(1 - \delta/2) < x < a(1 + \delta/2)$$
$$\psi = A_2 e^{-k_p(x - a - a\delta/2)} \quad \text{for} \quad x > a(1 + \delta/2), \tag{40}$$

where A_1, A_2 and B are constants to be determined by matching ψ and its derivatives at the boundaries. The functional dependence of ψ inside the ramp is chosen in accordance with Eq.(29); one can easily show that other terms in the expansion (29) can be neglected to linear order in δ. The plasma frequency at the resonant location

$$x_r = a\left(1 - \frac{\delta(k_p a - 1)}{2(k_p a + 1)}\right). \tag{41}$$

is equal to the *unperturbed* channel frequency, given by Eq.(20).

Using $\xi = x - x_r$ instead of x, we find the coordinates of the inner and outer edges of the ramp:

$$\xi_{in} = -\frac{a\delta}{1 + k_p a}$$

$$\xi_{out} = \frac{k_p a^2 \delta}{1 + k_p a}. \qquad (42)$$

By matching the derivatives at ξ_{in} and ξ_{out} we find, to lowest order in δ,

$$B = -\frac{1 + k_p a}{2a^2 \delta}, \qquad (43)$$

and, using Eq.(43) and continuity of ψ, recover

$$A_1 = 1 - \frac{\delta}{2(1 + k_p a)}$$
$$A_2 = 1 - \frac{\delta k_p^2 a^2}{2(1 + k_p a)}. \qquad (44)$$

Expressing the plasma quantities inside the density ramp (using $\omega = \omega_{ch}$) in the new coordinates yields

$$\epsilon(\xi) = -\frac{(1 + k_p a)\xi}{a\delta}$$
$$k_p^2(\xi) = k_p^2 \frac{\xi - \xi_{in}}{a\delta}. \qquad (45)$$

Substituting Eq.(45) in Eq.(39) and integrating it across the ramp gives

$$\frac{-k_p A_2}{1 - \omega_p^2/\omega^2} - \frac{A_1}{a(1 - \delta/2)} = -\frac{k_p^2}{1 + k_p a} \int_{\xi_{in}}^{\xi_{out}} \frac{d\xi}{\xi}(1 - B\xi^2)(\xi - \xi_{in}). \qquad (46)$$

The integral on the RHS of Eq.(46) has a singularity at $\xi = 0$ and must be taken in the principal value sense.

After some algebra, assuming $\omega = \omega_{ch} + \Delta\omega$, we find, to linear order in δ,

$$\frac{\Delta\omega}{\omega_p} = \delta \frac{k_p^2 a^2 (1 - k_p^2 a^2 - 2k_p a \ln k_p a)}{4(1 + k_p a)^{7/2}}. \qquad (47)$$

The perturbed channel frequency is thus a function of both the width of the channel wall and the average width of the channel.

The dependence of the frequency shift on the channel width a and channel wall thickness δ was also found by solving numerically the eigenvalue Equation (39) and is shown in Fig. 1, where $\Delta\omega/\omega_p$ is plotted as function of δ for wall boundaries given by: (i) $x_{in} = a$, $x_{out} = a(1 + \delta)$; (ii) $x_{in} = a(1 - 0.5\delta)$, $x_{out} = a(1 + 0.5\delta)$; (iii) $x_{in} = a(1 - \delta)$, $x_{out} = a$. In all three cases $k_p a = 1$. Note that the frequency shift is, as expected, almost equal to zero in case (ii), since $\Delta\omega$, given by Eq.(47) vanishes for $k_p a = 1$. We believe that the

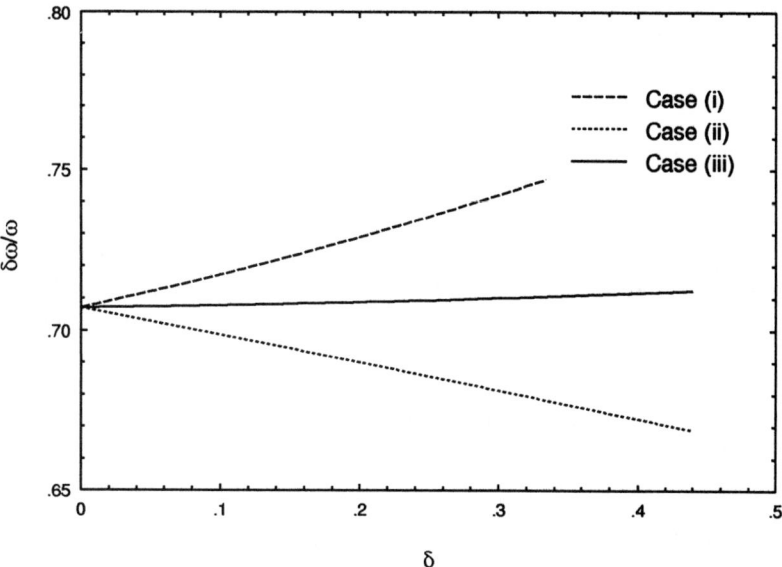

Figure 1: Normalized frequency shift as a function of a fractional wall thickness, for three different positions of the channel walls. In all cases $k_p a = 1$

frequency shift in case (ii) did not vanish identically due to a loss of numerical accuracy in the vicinity of the resonance point x_r. Case (i) gives a frequency shift almost identical to the shift that corresponds to making a step-function channel *wider* by a factor $(1 + \delta/2)$, while case (iii) corresponds to making a step-function channel *narrower* by a factor $(1 - \delta/2)$.

As Eq.(47) indicates, for typical parameters with $k_p a \approx 2$ the relative frequency shift of the wake mode due to a 10% wall thickness is about 1.0%. Hence, our intuition is correct and the (real) frequency of the wake excited inside a smooth wall channel is very robust to variations in thickness of the channel wall (as is the structure of the accelerating field inside the channel). Yet the gradual density ramp of the plasma introduces an imaginary part to the frequency of the channel. In the next section we calculate it perturbatively by evaluating the energy dissipated inside the density ramp per cycle of the wake oscillations.

4 Dissipation of the Wake

A simple observation helps to understand why the surface modes in the hollow plasma channel are dissipating. As we have observed in the previous section, there exists a location x_r inside the channel wall where the surface mode frequency matches the local plasma frequency. At this location the surface mode can mode-convert into the plasma wave. Since the plasma waves in the inhomogeneous plasma form a continuum, dissipation of the surface mode ensues. The effect of coupling to a continuous spectrum of modes was known for many years—one of the well-known application of this phenomenon is the Alfven heating, where a compressional (fast) Alfven wave dissipates by coupling to a continuum of shear (slow) Alfven waves [22, 23].

To calculate the rate at which the energy is dissipated inside one of the channel walls, the time averaged product of $\vec{E} \cdot \vec{j}$ is integrated over the volume of the wall:

$$\frac{dW}{dt} = \frac{b_z b_y}{2} Re \int_{a(1-\delta/2)}^{a(1-\delta/2)} dx \vec{j}^* \cdot \vec{E}, \qquad (48)$$

where b_z and b_y are the dimensions of the wake (which will cancel out when the *rate* of the decay is computed). Using the expressions Eq.(7) and Eq.(8) for the current and the electric field results in

$$\frac{dW}{dt} = -\frac{b_z b_y \omega_{ch}}{8\pi} Im \int_{a(1-\delta/2)}^{a(1-\delta/2)} dx \left(\frac{\partial P}{\partial x}\frac{\partial P^*}{\partial x} + \frac{\partial P}{\partial z}\frac{\partial P^*}{\partial z} \right) \frac{1}{1 - \omega_p^2(x)/(\omega_{ch} + i\nu)^2}, \qquad (49)$$

where ν is an infinitesimal positive quantity having the dimension of frequency, which is necessary to insure causality.

It is easy to convince oneself that the second term in the integrand of Eq.(49) dominates the first term in the limit of $\delta \ll 1$ (physically, this corresponds to the fact that the $x-$ component of the electric field is much larger at the resonant location than the $z-$ component; hence the dissipation due to $j_x E_x$ dominates the dissipation due to $j_z E_z$). Since the imaginary part of the integral in Eq.(49) comes from the singularity in the dielectric coefficient, the transverse dependence of P can be neglected and taken out of the integral. Using $\xi = x - x_r$ instead of x results in

$$\frac{dW}{dt} = -\frac{b_z b_y \omega_{ch}^3 \mid A \mid^2}{8\pi c^2} Im \int_{\xi_{in}}^{\xi_{out}} d\xi \frac{1}{i\tau - (1 + k_p a)\xi/(a\delta)}, \quad (50)$$

where τ is an infinitesimal positive number. Evaluating Eq.(50) yields

$$\frac{dW}{dt} = \frac{\mid A \mid^2 \omega_{ch} k_p^2 b_z b_y a \delta}{8(1 + k_p a)^2}. \quad (51)$$

The source of the energy dissipated in the channel walls is the stored energy of the wake, which can be easily calculated by summing the electric, magnetic, and particle kinetic energies inside the channel and in the bulk of the plasma. To zeroth order in δ this energy is given by

$$W = \frac{\mid A \mid^2 b_z b_y}{16\pi a}\left(1 + \frac{2(k_p a)^2}{3(1 + k_p a)} + \frac{k_p a}{2(1 + k_p a)}\left(1 + \frac{(2 + k_p a)^2}{(k_p a)^2}\right)\right). \quad (52)$$

We can now calculate the imaginary part of the surface mode as

$$\omega_{im} = -\frac{dW/dt}{2W}. \quad (53)$$

Substituting the rate of dissipation dW/dt from Eq.(51) and the wake energy W from Eq.(52) and evaluating Eq.(53) for $k_p a = 1$ we obtain:

$$\omega_{im} \approx 0.2\delta\omega_{ch}. \quad (54)$$

Hence, for $k_p a = 1$ and $\delta = 1.0$ one expects the amplitude of the wake to decay by about 70% after one oscillation.

An important question to examine is where this energy goes. Since no collisions were assumed to create this dissipation, one cannot attribute the loss of energy by the wake to bulk heating of the plasma. And, in fact, the singular fields that characterize the mode would take infinitely long to build up and would clearly violate the assumptions of our linear theory. In the next section we examine time evolution of the electric field inside the channel wall, excited by a ponderomotive force of an intense laser pulse.

The results of particle-in-cell (PIC) simulations [24] indicate that the wake is dissipated through the wavebreaking near the resonant location, and that there is production of energetic plasma electrons that take the energy out of the wake.

5 Time-Dependent Excitation of the Wake

After a detailed study of the properties of the channel eigenmodes in Section 3, we now examine how these wakes are excited by the ponderomotive force of the laser. To do that we rewrite Eq.(13) in the operator form:

$$\mathcal{L}P = -i\frac{\omega}{c}\tilde{f}\frac{\partial \ln \epsilon(x)}{\partial x}. \tag{55}$$

The function P can be expanded in a complete set of eigenmodes ψ_n of the operator $\mathcal{L}$ given by Eq.(36):

$$P(x,\omega) = \sum p_n(\omega)\psi_n(x,\omega) \tag{56}$$

One of the important conclusions of Section 3 is that the eigenfunctions ψ and the corresponding eigenvalues are only slightly perturbed by the finite thickness of the channel wall. Hence, to a high degree of accuracy, the ψ_n's can be chosen to be the eigenfunctions (and λ_n the eigenvalues) of Eq.(34) with a step-function density profile and $\omega_p^2(x)$ given by Eq.(18). Using the orthogonality conditions Eq.(37), the expansion coefficients can be expressed as

$$p_n = \frac{i\omega/c}{\lambda_n U_n}\int_0^{+\infty} dx'\tilde{f}(x',\omega)\psi_n(x',\omega)\frac{\partial}{\partial x'}\left(\frac{1}{\epsilon(x',\omega)}\right). \tag{57}$$

By examining the spectrum of Eq.(36) with a step-function density profile one can demonstrate that the only eigenvalue function $\lambda_n(\omega)$ that has a pole for *any* complex ω is the one with the discrete spectrum eigenvalues $\lambda_0(\omega)$, which vanishes at $\omega = \pm\omega_{ch}$. Alternatively, from the results of Section 2, there exists *only one* self-supported eigenmode (these self-supported modes correspond to poles of eigenvalue functions).

¿From Eqs.(8),(56) and (57) we can find the induced accelerating field inside the channel. Since $\lambda_0(\omega)$ has a pole at at $\omega = \pm\omega_{ch}$, we need to find its behavior in the vicinity of the pole. Since $\lambda_0(\omega)$ is regular in the vicinity of the pole, we consider $\omega > \omega_{ch}$, which, as we prove later, is equivalent to $\lambda > 0$. Solutions to Eq.(36) are then given by

$$\psi = \frac{\sin \lambda^{1/2}x}{\sin \lambda^{1/2}a} \quad \text{for} \quad x < a$$
$$\psi = e^{-(k_p^2-\lambda)^{1/2}(x-a)} \quad \text{for} \quad x > a, \tag{58}$$

where we have dropped the subscript on λ. Applying the continuity condition (17), with $f = 0$, yields an implicit $\lambda(\omega)$:

$$\frac{\tan\left(\lambda^{1/2}a\right)}{\lambda^{1/2}a} = \frac{\frac{\omega_p^2}{\omega^2}-1}{k_pa(k_p^2-\lambda)^{1/2}}. \tag{59}$$

Linearizing Eq.(59) in the vicinity of $\omega = \pm\omega_{ch}$ and $\lambda = 0$ yields:

$$\lambda = \frac{(1+k_p a)^2}{k_p a(-1/2 + k_p^2 a^2/3)} \frac{\omega^2 - \omega_{ch}^2}{c^2}. \tag{60}$$

We also compute the normalization constant

$$U_0 = \int_0^\infty \frac{dx}{\epsilon} \psi_0^2, \tag{61}$$

which, for $\omega = \pm\omega_{ch}$, is equal to

$$U_0 = \frac{-1/2 + k_p^2 a^2/3}{k_p^2 a}. \tag{62}$$

The accelerating gradient inside the channel is computed by observing that inside the channel $\omega_p^2(x) = 0$, so that the first term of Eq.(8) can be combined with Eq.(57) to obtain:

$$\begin{aligned} E_z(x,\zeta) &= \int \frac{d\omega}{2\pi c} e^{-i\frac{\omega}{c}(\zeta-\zeta')} \sum_n \frac{\psi'_n(x,\omega)}{U_n \lambda_n(\omega)} \\ &\times \int_{-\infty}^{+\infty} dx' \psi_n(x',\omega) \frac{\partial}{\partial x'} \left(\frac{1}{\epsilon(x',\omega)} \right) \frac{\partial f(x',\zeta')}{\partial \zeta'}, \end{aligned} \tag{63}$$

where the first integration is carried out along a contour in complex ω plane, above the real axis, to ensure causality. Even though this calculation appears to be much more cumbersome than the one used in Section 2, it is worthwhile to prove that Eq.(63) yields the same result as Eq.(24). The method of modal expansion, while almost being "too powerful" for finding E_z inside the channel, is the only method for finding the temporal evolution of the laser-driven fields inside the channel wall.

As mentioned earlier, only the lowest eigenmode of the discrete spectrum has an eigenvalue function $\lambda(\omega)$ with a pole. Thus, after closing the integration contour in the lower half of complex ω plane, summation in Eq.(63) contains only one non-vanishing term—the lowest discrete mode. Since the residues at the pole of $\lambda(\omega)$ at $\omega = \pm\omega_{ch}$ were computed earlier, we substitute Eqs.(60) and (62) into Eq.(63), and, adding the two residues, recover Eq.(24). Note that the second integral in Eq.(63) reflects the previously mentioned fact that the channel mode is driven by the *longitudinal* profile of the laser pulse at the plasma discontinuity (see Eq.(24)), where the derivative of the index of refraction is the largest.

We can now establish how the fields grow in time at the resonant location $x = x_r$. Formally, one expects the field to grow indefinitely since the field of the wake eigenmode diverges near resonant location (see Eq.(29)). An unusual physical situation occurs: as the laser pulse propagtes through the plasma, it generates a wakefield, which, near the resonant point, continues to grow *indefinitely* (in the linear theory, of course) *after* the laser pulse is gone!

As Eq.(8) indicates, the total electric field inside the plasma is given by a sum of two contributions. The first term on the RHS of Eq.(8), at a given location x, has a simple pole in the $\omega-$ plane, and, thus, cannot lead to a (formally) infinite growth of the field. Physically, the first term represents an oscillation, excited by the laser pulse, at the *local* plasma frequency $\omega_p(x)$. Thus, we can neglect this term and compute the contribution of the second term on the RHS of Eq.(8).

Similarly to Eq.(63), a modal expansion for E_x at an arbitrary location in plasma gives:

$$E_x(x,\zeta) = -\int \frac{d\omega}{2\pi c} e^{-i\frac{\omega}{c}(\zeta-\zeta')} \sum_n \frac{\psi_n(x,\omega)}{U_n \lambda_n(\omega) \epsilon(x,\omega)}$$
$$\times \int_{-\infty}^{+\infty} dx' \psi_n(x',\omega) \frac{\partial}{\partial x'}\left(\frac{1}{\epsilon(x',\omega)}\right) \frac{\partial^2 f(x',\zeta')}{\partial \zeta'^2}, \qquad (64)$$

As in the calculation of E_z, only an $n = 0$ mode can contribute to indefinite growth of the electric field. Other modes, while giving a nonzero contributions (unlike the case of E_z inside the channel), have an effect similar to that of the first term on the RHS of Eq.(8). (i.e., they have a simple pole at the local plasma frequency. By retaining only the $n = 0$ mode, and setting $x = x_r$, observe that the integrand in Eq.(64) has a double pole at $\omega = \pm\omega_{ch}$. We thus obtain:

$$E_x(\zeta, x_r) = \frac{k_p^2 a}{1 + k_p a} \int_{-\infty}^{\zeta} d\zeta' G_x(\zeta - \zeta') \frac{d^2 f(\zeta', x = a)}{d\zeta'^2}, \qquad (65)$$

where $\bar{f}$ is given by Eq.(23), and $G_x(\zeta - \zeta')$ is a Green function of the transverse electric field at resonant point given by

$$G_x(\zeta) = \frac{\zeta}{2}\cos(k_{ch}\zeta) + \frac{\sin(k_{ch}\zeta)}{k_{ch}}. \qquad (66)$$

The wake has the temporal behavior we expected—it grows indefinitely with time.

Similarly, one can estimate that the axial field at x_r will be growing logarithmically with time. If E_{x0} is the transverse field at the edge of the channel, the amplitude of the field at $x = x_r$ roughly grows as

$$E_x(x_r) \propto \frac{k_{ch}\zeta}{2} E_0. \qquad (67)$$

Thus, after the laser passes through the channel, electric field inside the ramp region continues to grow *in the absence* of an external driver.

The situation here is (at least formally) very similar to the excitation of a linear plasma profile by an external capacitor. Refs. [25]-[28] analyze an inhomogeneous plasma with a linear density ramp which is subject to a sinusoidally varying (with time) electric field produced by a plane capacitor. The electric field is co-linear with the density gradient and its frequency matches the local plasma frequency at the resonant location $x = x_r$. An obliquely incident $p-$ polarized electromagnetic wave can also play the role of a capacitor. Secular growth of the electric field at resonant point, which saturates through collisions, relativistic corrections, wavebreaking, particle ejection, etc., was predicted [25, 26]. The important difference between the hollow channel and the excitation of an inhomogeneous plasma with a capacitor is that *no external capacitor* is needed in our case. Instead, a hollow channel mode is excited, which serves *as a capacitor* to drive the internal modes of inhomogeneous plasma.

Initially the fields *inside* the channel and in the inhomogeneous plasma region are not influenced by the dramatic growth of the field at resonant point. But later, as E_x builds up, wavebreaking [29] occurs, energetic electrons are ejected into the channel, thereby dissipating the wake. The process of wavebreaking and electron acceleration (in the direction of decreasing plasma density, that is, into the channel) has been extensively discussed in the literature [25]-[28].

The mechanism for wave breaking in the limit of $r_0 \ll a\delta$ was discussed in the literature [25, 28]. Briefly, the spatio-temporal structure of the field near resonance point is such that it has the form of a wave packet, moving with a characteristic phase velocity $v_{ph} \approx \omega_{ch} \Delta L$, where ΔL is the size of a resonance region. Wave breaking occurs when electron velocity (which grows linearly with time) matches v_{ph} (which decreases linearly with time). The number of oscillations before the wave breaks is roughly given by

$$N = \omega_{ch} t \simeq (a\delta/r_0)^{1/2} \qquad (68)$$

(see [27]). In the case of $r_0/(a\delta) = 100$ wavebreaking was numerically observed [24] to occur after the first 1.5 oscillations, in good agreement with Eq.(68).

It is important to emphasize that it is not the wavebreaking inside the channel wall *per se* which causes the decay of the accelerating wake. Wavebreaking is only a mechanism for transferring the energy from the wake to the energetic electrons, and the rate at which this transfer occurs clearly depends on the intensity of the electric field at the resonant location. The effective Q of the plasma channel is given by

$$Q = \frac{\omega_{ch}}{\omega_{im}} \qquad (69)$$

, where ω_{im} is the imaginary part of the mode given by Eq.(53) and quantified for a particular case of $k_p a = 1$ by Eq.(54).

Dissipation of the accelerating mode imposes serious limitations on the operation of future laser wakefield accelerators. It introduces an effective Q of the plasma channel, which can be quite low [24] and prevents accelerating multiple bunches by the wake created in a single laser shot. It appears that to increase Q one has to use channels with a sharp boundary, as suggested by Eq.(54). Nonlinear dissipation of the wake in a channel with $a\delta \succeq r_0$ is being presently investigated.

6 Conclusions

In this paper we have analyzed the problem of wake excitation in an inhomogeneous plasma. A general formalism for excitation of the accelerating mode in nonuniform plasma was developed and a hollow channel accelerator with a step-function density profile [1] was considered as an example. The eigenfrequency of the accelerating mode was derived and the excellent properties of the mode for particle acceleration (uniformity in the transverse dimension) were derived using this formalism. The time-dependent excitation of the mode by the ponderomotive force of a laser pulse was derived and the accelerating gradient inside the channel was expressed as a causal time convolution of the ponderomotive force at the edge of the channel and the wake Green function.

For the first time a more realistic case of a channel with a smooth interface between vacuum and plasma was considered. We have proven that the structure of the accelerating mode is only slightly perturbed by the finite thickness of the channel walls. The corrections to the eigenfrequency of the accelerating mode due to finite thickness of the channel was obtained analytically and numerically. The interaction between the channel mode and the inhomogenious plasma inside the channel was considered. It was shown analytically that the transverse electric field at the particular location inside the channel wall, where the local plasma frequency matches the channel eigenfrequency, is driven resonantly. A Green function for the transverse electric field at the resonant location was derived, predicting a linear growth of the amplitude of the field with time.

A somewhat exotic scenario was predicted, in which the transverse electric field at resonant location grows indefinitely (in linear theory) *after* the short laser pulse (which excited the wake) has already propagated downstream. Nonlinear effects, most notably wavebreaking, which arrest the growth of the electric field, were briefly discussed. The important implication of these effects is that the accelerating wake is dissipated, thereby introducing an effective quality factor Q for the channel accelerator and limiting the number of electron bunches that can be accelerated in the wake of a single laser pulse. An important continuation of this work will be to carry out PIC simulations for different thicknesses of the channel wall (much larger or much smaller than the excursion of a surface electron, etc.) to identify the most promising regime of operating the hollow channel laser wakefield accelerator.

References

[1] T. Katsouleas, T. C. Chiou, C. Decker, W. B. Mori, J. S. Wurtele, G. Shvets, and J. J. Su, in *Advanced Accelerator Concepts*, edited by J. S. Wurtele, (AIP, New York, 1993), p. 480; T. C. Chiou, T. Katsouleas, C. Decker, W. B. Mori, J. S. Wurtele, G. Shvets, J.J. Su, Phys. Plasmas **2**, 310 (1995).

[2] T. Tajima and J. M. Dawson, *Phys. Rev. Lett.* **43**, 267 (1979); L. M. Gorbunov and V. I. Kirsanov, Zh. Exp. Teor. Fiz. **93**, 509 (1987) [Sov. Phys. JETP **66**, 290 (1987)]; C. Joshi, W. B. Mori, T. Katsouleas, *et al.*, Nature, **311**, 525 (1984).

[3] J. S. Wurtele, Phys. Fluids B **5**, 2363 (1993); J. S. Wurtele, Physics Today **47** (no. 7), 33 (1994), and references therein.

[4] P. Sprangle, E. Esarey, Phys. Fluids B **4**, 2241 (1992).

[5] C.E. Clayton, et al., Phys. Rev. Lett. **30**, 37 (1993).

[6] C. G. Durfee III, H. M. Milchberg, Phys. Rev. Lett. **71**, 2409 (1993); C. G. Durfee III, J. Lynch, H. M. Milchberg, Opt. Lett. **19**, 1937 (1994).

[7] P. Sprangle, E. Esarey, and A. Ting, *Phys. Rev. Lett.* **64**, 2011 (1990).

[8] C. E. Max, J. Arons, and A. B. Langdon, *Phys. Rev. Lett.* **33**, 209 (1974).

[9] G. Schmidt and W. Horton, *Comments Plasma Phys. Controlled Fusion* **9**, 85 (1985).

[10] G. Z. Sun, E. Ott, Y. C. Lee, and P. Guzdar, Phys. Fluids **30**, 526 (1987).

[11] W. B. Mori, C. Joshi, J. M. Dawson, D. W. Forslund and J. M. Kindel Phys. Rev. Lett. **60**, 1298 (1988).

[12] T. M. Antonsen, Jr.,and P. Mora, *Phys. Rev. Lett.* **69**, 2204 (1992); T. M. Antonsen, Jr., and P. Mora, Phys. Fluids B **5**, 1440 (1993).

[13] C. J. McKinstrie and R. Bingham, Phys. Fluids B **4**, 2626 (1992).

[14] C. Decker, W. B. Mori, and T. Katsouleas, *Phys. Rev. Rap. Comm.*, in press 1994.

[15] P. Sprangle, E. Esarey, J. Krall, and G. Joyce, *Phys. Rev. Lett.* **69**, 2200 (1992).

[16] G. Shvets and J. S. Wurtele, Phys. Rev. Lett., **73**, 3540 (1994).

[17] P. Sprangle, J. Krall, and E. Esarey, Phys. Rev. Lett. **73**, 3544 (1994).

[18] G. Shvets, PhD thesis, MIT (1995) (unpublished).

[19] J. P. Freidberg, R. W. Mitchell, R. L. Morse, and L. I. Rudsinski, Phys. Rev. Lett. **28**, 795 (1972).

[20] V. L. Ginsburg, Propagation of Electromagnetic Waves in Plasma, Addison-Wesley, Reading, Mass. (1964).

[21] W. L. Kruer, The physics of laser plasma interactions (Addison-Wesley, 1988), and references therein.

[22] R. Cross, An Introduction to Alfven Waves (Adam Hilger, Bristol, 1988), and references therein.

[23] Chen L. and Hasegawa A., Phys. Fluids **17**, 1399 (1974).

[24] T. Katsouleas, Private Communication.

[25] J. Allbritton and P. Koch, Phys. Fluids **18**, 1136 (1975).

[26] L. M. Kovrizhnykh and A. S. Sakharov, Sov. L. Plasma Phys. **5**, 470 (1979).

[27] P. Koch, Phys. Fluids **16**, 651 (1973).

[28] S. V. Bulanov, L. M. Kovrizhnykh, and A. S. Sakharov, Physics Reports, **186**, 1 (1990).

[29] J. M. Dawson, *Phys. Rev.*, vol. 113, p.383, 1959.

Collective Instabilities in Accelerator and Storage Rings

C. Pellegrini

UCLA, Physics Department
405 Hilgard Avenue, Los Angeles, CA 90095

Abstract We describe the initial discovery of collective beam instability in accelerator and storage rings, the work that was done in the 60's to obtain a theory of these effects, and their impact on the developments of electron-positron and proton-proton colliders.

1. Introduction

Preparing this contribution to Andy Sessler's Symposium has been a way to look back in time at accelerator and collider development. Looking back in time we can see:

1930-1950, early times: developments of electrostatic accelerators, linear accelerators, cyclotrons.

1950s, synchrotrons, early ideas on colliders; introduction of strong focusing, and studies of single particle dynamics, resonances, tolerances;

1960s, the early collider period; the decade of collective beam instabilities: realization that a beam is an essentially unstable system;

1970s, the mature colliders period: larger energies, larger colliders, more luminosity;

1980s-90s, the very large colliders period.

Looking to the future, to the XXI century, we see the continued need to reach higher energy and luminosity, while keeping the size and cost of the colliders within affordable limits. We still do not know how these goals will be reached, and we can only stress the importance of new innovative ideas, and of supporting the work being done, by Andy Sessler and other scientists, in this direction.

In this paper I will mainly look at what happened during the past, and in particular to the time between the late 1950s, and the early 1970s, the decade of the instabilities. Andy Sessler has been a leading figure in shaping our view of the collective behavior of a beam. I started to work in this field in the early 1960s, and I was lucky enough to be able to work with and learn from Andy Sessler, Bruno Touschek, Matthew Sands. I will try to describe some of the work and the excitement of that period, discussing in particular the discovery and the development of our understanding of the Negative Mass Instability, the Resistive Wall Instability for coasting and bunched beams, the Bunch Lengthening and Potential Well Distortion Effect. This by no means exhausts the wealth of work done at that time, but can hopefully give the reader the flavor of the problems encountered, and the way they were solved. A review of the work done on collective instabilities and other aspects of the collider development can be found in ref. 1.

2. Collective Instabilities in a Storage Ring.

The work on instabilities started in the MURA group. The existence of instabilities in particle beam was discovered theoretically by Nielsen, Sessler and Symon (2, 3), who studied the negative mass instability.

Their results were presented at the International Conference on High Energy Accelerators at CERN in 1959. . The basic mechanism of the negative mass instability can be described by the following steps:

1. Particles moving on a circle in an accelerator can have a frequency decreasing with energy; if you make positive work on them their frequency will decrease;

2. If I sit on one electron in a ring and I am pushing back another electron following me -through the Coulomb force- the electron energy will decrease, its frequency will increase and the electron will come nearer to me!

3. The result is that if I have initially a coasting beam, with uniform longitudinal density distribution, any longitudinal density perturbation will lead to bunching and to an increase in energy spread.

Beam Instabilities where one of the main concerns for colliders during the sixties. Transverse coherent oscillations of single beams were observed in the early sixties at MURA, Brookhaven, CERN, Argonne. The first beam tests on the 500 MeV Princeton-Stanford ring showed the existence of a vertical instability. The theory of the Resistive Wall Instability was developed in 1963 by Laslett, Neil and Sessler (4) to explain these observations. The theory showed that when a charged particle moves near to a resistive wall, the current driven in the wall decays slowly, and produces an electromagnetic field that drives the oscillations of the following particles. In a ring a particle can also "follow" itself after a revolution, and thus become unstable.

Longitudinal coherent instabilities were also observed at CERN, MURA and Saclay. The instability of a coasting beam can be interpreted with the negative mass instability, driven by space-charge forces, and/or the longitudinal resistive wall instability (2, 5), driven by the currents in the wall. In the case of a bunched beam the bunches interact also with the Radio Frequency accelerating cavity. Since the fields generated in the high Q cavity by one bunch can also last a long time, the same fields can produce a force on a following bunch, and again the situation can lead to an instability. This effect was analyzed by K. Robinson (6) and Auslender (7).

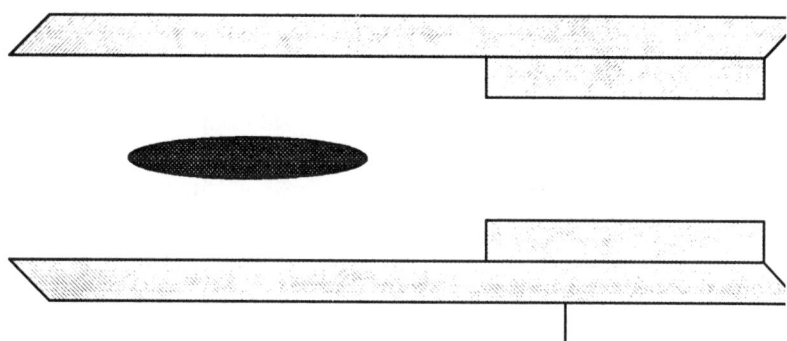

FIGURE 1. A particle bunch travelling through a piece of vacuum chamber with a discontinuity.

This was the situation in the summer of 1965. Several storage rings were already in operation: the 500 MeV electron-electron Princeton-Stanford ring at

Stanford; the 200 MeV electron-positron AdA at Frascati and Orsay; the 130 MeV electron-electron ring and the 700 MeV electron-positron ring at Novosibirsk. Bremsstrahlung produced in the electron-positron interaction had been observed on AdA, and electron-electron wide angle scattering had been observed in the Princeton-Stanford ring. New electron-positron rings, ACO, Adone, Spear, with larger energies and luminosity were being designed at Orsay, Frascati and Stanford. A 30 GeV proton-proton ring, the ISR was being designed at CERN.

The coherent instabilities, and the beam-beam interaction, were a main concern for these projects. The scientists designing the colliders believed to have a good understanding of the instabilities observed until that time, but there was a concern that something new might happen going to larger energies and intensities. There was also a need to extend the existing theories to cases which had not been studied, as for example the transverse resistive wall instability of bunched beams. A workshop was organized at SLAC in the summer of 1965, by Richter, Sands and Sessler to " bring together the interested workers for a concentrated attack on the outstanding problems " (8).

The workshop was held at SLAC from June 28 to July 30, 1965, with 19 participants. I remember it as a very exciting event. When the workshop started the main questions that were asked were:

1. What is the effect of the Resistive Wall Instability for a beam with many short bunches?

2. In what ways are the effects in an electron-electron ring and an electron-positron ring similar or different?

3. are colliding beam stable generally with respect to coherent longitudinal (phase) oscillations?

4. Is the colliding beam incoherent stability sufficiently well understood (on the basis of the experience of the electron-electron rings and the semiheuristic calculations of Courant and others) to warrant extrapolation to a 3 GeV machine with current as high as 1 Ampere?

5. are there any effects not thought of yet- particularly effects which might only show up at ultrarelativistic energies?

Some of these questions were answered at the Workshop, as can be seen from reading the summary talk given by Andy Sessler (8), and additional work followed in the next few months. In particular much attention was given to understanding the resistive wall instability of a bunched beam. The wake fields behind a charge traveling in a resistive cylindrical pipe were analyzed in detail by many people, showing that they can persist for a very long time, in fact for a transversely oscillating charge they decay only as the inverse square root of the time.

An analysis of the dynamics of a bunched beam was also done by Courant and Sessler (9) and other authors (10). This work showed that in a beam with B equal bunches there are B normal modes of oscillations about half of which are unstable in the absence of Landau damping. Hence the beam is essentially unstable, and either Landau damping, or radiation damping, or a feedback system must be used to obtain stability for a non zero current.

The theory of the resistive wall instability was the tool being used to design the next generation of colliders, and this was justified by the agreement between the observations at the Cosmotron, AGS, Stanford and other rings. There was however a discrepancy between the theory and experimental observation at MURA, and this discrepancy was investigated by Laslett, and by Briggs, Neil and Sessler, continuing previous work by Chirikov at Novosibirsk. This work showed that the real and imaginary part of the frequency shift due to the interaction with the walls can be very sensitive to a number of effects, like discontinuities in the vacuum chamber wall, as shown in Figure 1, corrugations in the longitudinal direction (bellows), clearing electrodes, trapped ions, or curvature of the particle orbit or vacuum tank.

Clearings electrodes seemed to be responsible for the discrepancy in the MURA case, and it was shown that they, and also other elements, can change the instability threshold by more than one order of magnitude. This opened the way to a much more general description of the interaction of the beam with its surrounding environment, in which instead of using an idealized cylindrical or rectangular vacuum chamber one includes a more detailed description, paying particular attention to elements that can resonate at particular frequencies, or that can reduce the cancellation between magnetic and electric forces in the term producing the real part of the frequency shift. In the end this led to what is now called the "impedance budget", an important part of any storage ring design and construction.

Another area of concern and study was the coherent instabilities of two counter rotating beams in a collider. Preliminary work had been done by Sessler in 1964, and by Ritson and Rees prior to the Workshop. The work was extended by Pellegrini and Sessler and further extended and published by them in 1967 (11). The result was very simple: for two beams with the same equilibrium orbit and equal bunches, the single beam modes of oscillations are unchanged, except for the zero (equal bunch phase) modes which are coupled.

Other problems discussed and reported in (8) were the longitudinal instability of bunched beams (Robinson), and the beam-beam interaction for beams with many bunches, and for protons beams. For electron-positrons colliders the case of interest was that of two beams with many bunches, where each bunch can interact with many bunches of the other beam, with some of these interactions taking place in points different from the ones were we observe the collisions (Amman). In the case of proton storage rings the problem was that of short term beam blow up (Keil), and possible long term, secular instabilities due the periodic nonlinear force in the beam-beam interaction (Hine).

In his summary talk Sessler presented at the end a long list of still unsolved problems, and of areas where further work was needed and were surprises could be expected. Some of the points that he raised are:

1. Coherent instabilities of a single beam:
    1.1 influence of various wall materials and types and their influence on the $(1-\beta^2)$ cancellation in the force; influence of clearing electrodes;
    1.2 studies of nonlinear phenomena and the effect of ions
2. Coherent instability in a single bunched beam:
    2.1 Are the very high modes corresponding to the internal motion of a bunch stable?
3. Coherent instabilities involving two beams:
    3.1 Extend the analysis to bunched beams

It is interesting to notice that question 2.1 was answered by the discovery and analysis of the head-tail instability, that question 3.1 led to an analysis of an important mechanism for the beam-beam interaction, that the effect of ions, question 1.2, as been a major source of problems for many rings, and that the problem raised in 1.1 has led to what we call today impedance budget, a major element in any ring design.

3. Other Instabilities and Conclusions

Other instabilities that received attention where the bunch lengthening and energy spread increase, that were observed in ACO, Adone and the Novosibirsk rings. Part of the explanation for this effect was the distortion in the RF potential well (12), and partly a high frequency negative mass type instability leading to turbulence within the bunch (13).

The study of collective beam instabilities in storage rings has led us to develop a good understanding of the physics of these systems, and this understanding has helped us in other areas. An example of this larger view of instabilities is the Free Electron Laser, another system on which Sessler has worked, giving important contributions which will be illustrated by William Barletta in his talk.

In its simplest formulation the FEL can be described as a set of electrons interacting with a monochromatic electromagnetic radiation field, while moving in an undulator magnet. When the beam has a uniform longitudinal density distribution, as is usually the case when the beam enters the undulator, the state of the system is not an equilibrium state. Hence an instability will develop that will move the state of the system toward an equilibrium state, in which the longitudinal density distribution is bunched at the radiation wavelength. The radiation emitted from this bunched beam, at the frequency corresponding to the bunching, is then enhanced. This instability is very similar to the generalized negative mass instability, the microwave instability, of a storage rings, and it can be described using similar techniques.

To conclude I want to say again that the work on collective instabilities, important as it has been, is only part of Andy Sessler's contribution to the physics of particle beams. As he has been working at MURA when all of this started, he is now one of the leaders in producing and studying novel ideas for the colliders of the next century, and for other applications of particle beams.

References

1. The Development of Colliders, C. Pellegrini and A. M. Sessler eds., American Institute of Physics Press, New York 1995.

2. C.E. Nielsen and A.M. Sessler, Rev. Sci. Instr. 30, 80 (1959).

3. C.E. Nielsen, A.M. Sessler and K.R. Symon in Proc. Intern. Conf. on High Energy Accelerators, CERN, 239 (1959).

4. L.J. Laslett, V.K. Neil and A.M. Sessler, Lawrence Radiation Laboratory Report UCRL 11090 (1963) and Rev. Sci. Instr. 36, 436 (1965).

5. V. K. Neil and A.M. Sessler, Rev. Sci. Instr. 36, 429 (1965).

6. K. W. Robinson, Stability of Beams in Radio Frequency Systems, Cambridge Electron Accelerator Intern. Rep. CEAL-1010 (1964);

7. V. L. Auslender et al., Phase Instability of Intense Electron Beams in a Storage Ring, Proc. Inter. Conf. on High Energy Accelerators, p. 339, Rome (1965).

8. 5. B. Richter, M. Sands, and A.M. Sessler, eds. Storage Ring Summer Study on Instabilities in Stored Particle Beams SLAC Rep. 49 (1965).

9. E.D. Courant and A.M. Sessler, Rev. Sci. Instr. 37, 1579 (1966).

10. E. Ferlenghi, C. Pellegrini and B. Touschek, Nuovo Cimento 44B, 253 (1966).

11. C. Pellegrini and A.M. Sessler, Proc. Inter. Conf. on High Energy Accelerators, Cambridge, Mass. (1967).

12. C. Pellegrini and A.M. Sessler, Nuovo Cimento 3A, 116 (1971).

13. P. J. Channel, and A.M. Sessler, Strong Turbulence and the Anomalous Length of stored Particles Beams, Nucl. Instr. and Meth. 136, 473-484 (1976).

MAXIMIZING THE LUMINOSITY OF ELOISATRON, A HADRON SUPERCOLLIDER AT 100 TeV PER BEAM

William A. Barletta

Lawrence Livermore National Laboratory
Livermore, CA 94610
and
Department of Physics
University of California Los Angeles

INTRODUCTION

In general as one raises the energy of a collider one must simultaneously increase the luminosity to compensate for decreasing cross sections. Applying the presently available accelerator technology embodied in the designs of the LHC (8 TeV per beam at 10^{34} cm^{-2} s^{-1}) and the SSC (20 TeV per beam at 10^{33} cm^{-2} s^{-1}) to the ELOISATRON, a proton collider operating at 100 TeV per beam yields a collider design with a luminosity of 10^{34} cm^{-2} s^{-1}. To extend the physics reach of super-colliders to the maximum possible extent will require designing the collider to achieve luminosities $> 10^{35}$ cm^{-2} s^{-1}.

This paper presents a general context for assessing the performance of hadron supercolliders in general and ELOISATRON in particular. It begins with an illustration of machine trends and with definitions of key collider characteristics. A brief description of design strategies to accommodate limiting beam physics and limiting technologies constitutes the next section. A simple spreadsheet based computer code incorporating these considerations allows one to perform self-consistent parameter searches to yield design characteristics of ELOISATRON (ELN) with maximum possible luminosity. To underscore the point that near term technology is applicable to the limits of the high energy frontier, the paper concludes with a sketch of the characteristics of the Ultimate ELOISATRON (UELN), a hypothetical collider of higher energy than the ELOISATRON.

COLLIDER TRENDS IN ENERGY AND LUMINOSITY

The search for understanding the nature of mass and the dynamics underlying the physical universe have led particle physicists to seek to build colliders with ever higher beam energies and luminosities. The performance trends in present and future hadron colliders are illustrated in figure 1. The limits on collider performance are determined both by beam physics and by available technology. This paper offers a frame-work for exploring the systematics and in particular the energy dependence of limiting beam physics and limiting technologies.

© 1996 American Institute of Physics

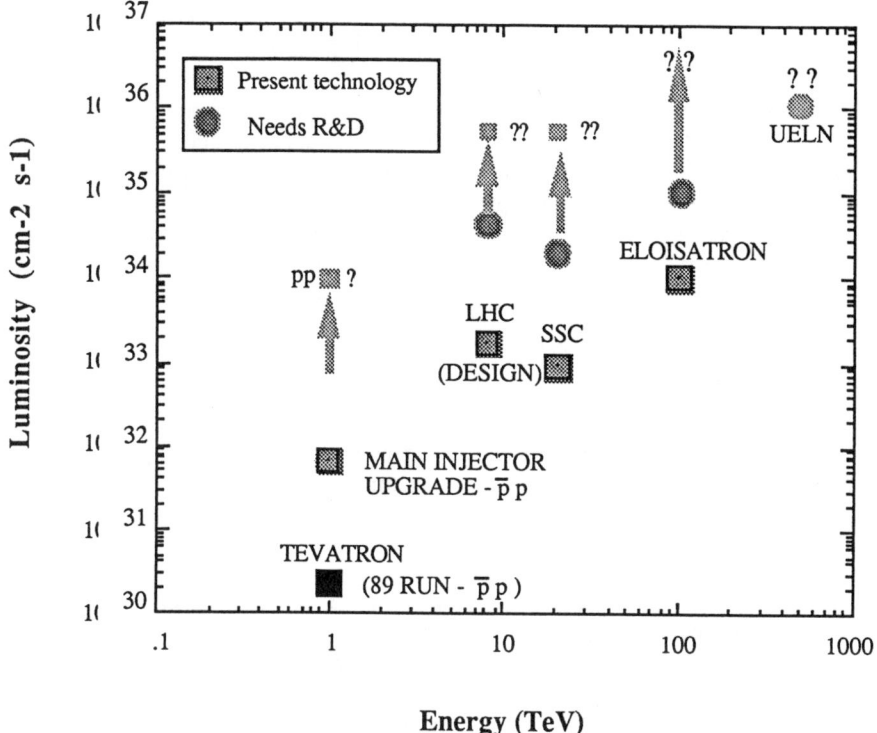

Figure 1. The luminosity goals of present and future hadron colliders

<u>Luminosity</u>

For bunches of equal population, N, and equal sizes at the interaction point colliding at a frequency f_{coll}, the luminosity is given by the well known expression

$$\mathcal{L} = \frac{N_1 N_2 f_{coll}}{4 \pi \sigma_x \sigma_y (1 + q^2)^{1/2}} \quad (1a)$$

where σ_x and σ_y are the Gaussian horizontal and vertical radii and where q accounts for the luminosity loss due to a crossing angle, α in the vertical plane;

$$q = \frac{\alpha \sigma_z}{2 \sigma_y}. \quad (1b)$$

Writing the beam radius in terms of the normalized emittance, ε_n, the relativistic factor, γ, and the value of the β-function at the interaction point, β^*, and ignoring the effects of a non-zero crossing angle, one has

$$\mathcal{L} = \frac{N^2 c \gamma}{4\pi \varepsilon_n \beta^* S_B} = \frac{1}{e\, r_p}\left(\frac{N\, r_p}{4\pi \varepsilon_n}\right)\left(\frac{\gamma I}{\beta^*}\right) \equiv \frac{1}{e\, r_p}\,\xi\!\left(\frac{\gamma I}{\beta^*}\right), \qquad (2)$$

where I is the average beam current and r_p is the classical radius of the proton. The quantity ξ is the linear (head-on) tune shift produced by the beam-beam interaction. Eq. (2) displays a natural linear growth of the luminosity with beam energy. The "pain" associated with increasing the luminosity faster than the natural linear growth of luminosity with energy derives from the necessity to increase the beam current simultaneously with increasing γ. How should one choose ε_n, β^*, S_B, and N as a function of the beam energy to maximize the luminosity? What constrains the choices?

DESIGN STRATEGIES

A recent look at maximizing the luminosity of the SSC and the LHC (Snowmass, 1990) suggested the following approaches to collider design:

1) increasing the charge per bunch,
2) increasing the number of bunches,
3) increasing the crossing angle to allow more rapid bunch separation
   thereby reducing parasitic crossings,
4) tilting the bunch with respect to the direction of motion at the interaction point ("crab-crossing"),
5) minimizing the β function at the interaction point.

These strategies together with their attendant physics and technology issues are exactly those being pursued in design studies of very high luminosity B and φ factories at SLAC, CERN, Cornell, KEK, UCLA, and Laboratorio Nazionale Frascati (LNF). One of the major technical difficulties in raising the luminosity in proposed flavor factories from the level achieved in CESR (2×10^{32} cm^{-2}s^{-1}) to that required to explore the nature of CP violation is handling the intense synchrotron radiation that is generated by the multi-ampere beam currents.

Upon close examination of the challenge of reaching the highest possible luminosity both in ELOISATRON and in lower energy hadron supercolliders, one finds that the physical phenomenon that underlies nearly all design difficulties in the range from 10 to 100 TeV (and beyond) is the emission of synchrotron radiation by the protons. Even the practical difficulties of controlling the consequences of synchrotron radiation are similar to those for electron rings if one assumes (as is generally the case) that the vacuum pipe in the hadron super-collider must operate at cryogenic temperatures. If, however, the operating temperature of the new, high T_c superconductors can be pushed to ≈ 300 °K, very high electrical conductivity need not imply cryogenic temperatures. In that case even at 100 TeV the synchrotron radiation load from the high current proton beam will easily be within the range routinely handled in existing electron storage rings. In that case the ultimate energy per beam would not be limited to 100 TeV by synchrotron radiation; instead, a PeV collider could be considered.

Accelerator physicists generally expect that the design approaches listed above should achieve the desired performance goals. Nonetheless, although an ELOISATRON with a luminosity of 10^{34} cm^{-2} s^{-1} may be quite conventional with respect to its constituent technologies, achieving the highest possible luminosity at 100 TeV (>10^{36} cm^{-2} s^{-1}) will push both beam characteristics and accelerator technologies considerably beyond present practice. Nonetheless, the required levels of performance are not beyond reasonable extrapolations of state-of-the-art accelerator technologies.

Here, the skeptic might object that regardless of its design, a collider of conventional technologies with an energy five times higher energy than that of the SSC would cost roughly five times as much as the SSC, that such a cost is unacceptably high, and hence that conventional designs are not economically practical. Fortunately, in the context of an appropriate partnership between scientific laboratories and industry that maximizes the use of existing technical and manufacturing infra-structure, this crude economic argument does not have force. The costs of the SSC project are strongly related to the style of its execution and its construction which required the established of a new, large-scale technical infra-structure. This choice has certainly received substantial criticism in technical circles.

From a technological point of view the cost of any hadron supercollider will depend strongly on the ultimate luminosity for which it is designed. Economic considerations notwithstanding, to achieve the highest possible luminosities (>10^{36} cm^{-2} s^{-1}), existing technologies must be pushed into new regimes (e.g., by finding practical, high T_c super-conductors suitable for magnet windings). In that case the design strategies must receive detailed *experimental* exploration for one to arrive at level of confidence commensurate with the cost of a 100 TeV collider with a luminosity in the range of 10^{36} cm^{-2} s^{-1}.

The beam dynamics limiting luminosity are embodied in the maximum tune shift that can be achieved during collisions. In existing hadron colliders such as the Tevatron and the Sp$\bar{p}$S where radiation effects are negligible, the total tune shift from all of the multiple interaction points is found to be 0.024. As it is unclear whether such a large tune shift can be realized from a single interaction point, the Snowmass study took 0.01 as a limiting value. In the ELOISATRON operating at ≈100 TeV, radiation damping will become an important determinant of the radial distribution of the beam. In the absence of dilution effects to the collisions and feedback control of instabilities, this distribution would resemble that seen in electron storage rings. Then the maximum tune shift in a collider with a single interaction point may no longer be limited to 0.01, but might approach the value of 0.03 (or even 0.06) that has been achieved in high energy electron-positron colliders (PEP at SLAC). A more pessimistic scenario is that fast dilution plus radiation damping might actual broaden the beam distribution vis á vis that in electron rings. Then the tune shift may actually be lowered.

One of the best ways to study the limits of stability in any system is to make frequent excursions into the unstable operating regime. Unfortunately the consequent loss of beam would impermissible in existing storage rings that must also provide high integrated luminosity to users. To develop the requisite experimental database one might instead consider using a low energy e$^+$–e$^-$ collider operating at ≈100 MeV per beam with high currents and with several bunches. With a flexible lattice such a storage ring would permit both high and low emittance tunes. With the installation of radio frequency deflection cavities ("crab" cavities), the sensitivity of crab-crossing to synchro-betatron resonances can be tested. These features are similar to those of DAΦNE, the φ factory under construction at the Laboratorio Nazionale Frascati.

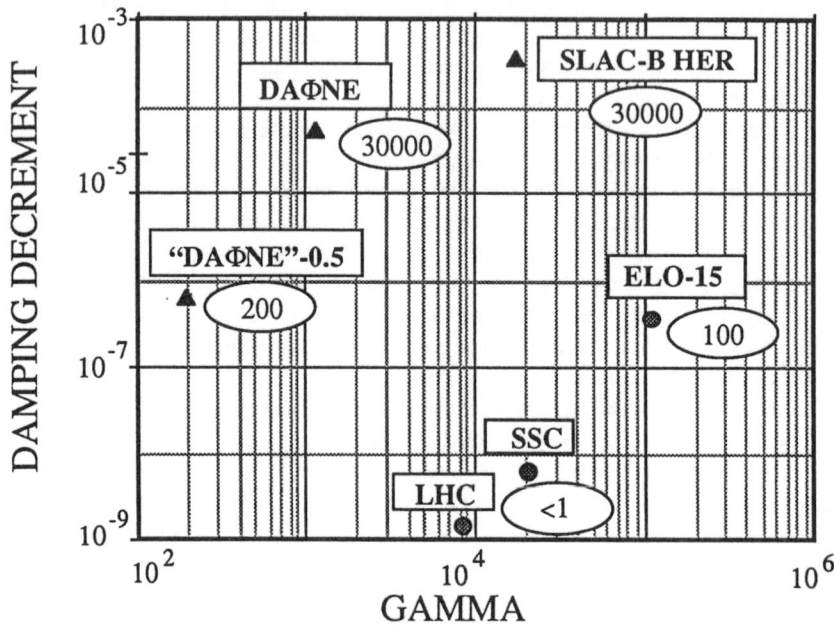

Figure 2. Radiation damping characteristics of colliders

The radiation damping characteristics of e+–e- colliders and hadron super-collider are shown in Fig. 2. The number encircled is the number of damping times per luminosity lifetime at 10^{34} cm^{-2}s^{-1} luminosity.

Designing for maximum luminosity yields a collider with a very large number bunches that implies beam crossing rates of the order of 0.1 — 1 GHz with several tens of collisions per crossing. Determining whether detectors and data acquisition and processing systems can be designed to accommodate the enormous rates implied by extremely high luminosities needs a serious evaluation by a group of detector specialists working with accelerator physicists familiar with hadron and e+-e- colliders.

The implications of trying to maximize luminosity of ELN in a self-consistent design can be explored most easily using a computer code to perform parameter searches. In the absence of experimental evidence of higher values of the maximum head-on tune shift per interaction point, one should take this value to be 0.01 as used in the Snowmass study. With all the cautions stated above, on the basis of such a parameter exploration it seems practical to construct a 100 TeV per beam collider with a luminosity $>10^{34}$ cm^{-2} s^{-1} using the same technologies being realized for the LHC. At energies below its top value such an ELOISATRON would have an energy vs.

luminosity dependence as shown in Figure 3. Reducing the allowable radiation to 2 W/m would reduce the luminosity at 100 TeV by a factor of two. In contrast, raising the value to 20 W/m yields 10^{35} cm^{-2}s^{-1}. Parameters for both of these cases are given in Tables 1 and 2 respectively. In neither case does the use of crab-crossing seem to offer much increase in the design luminosity.

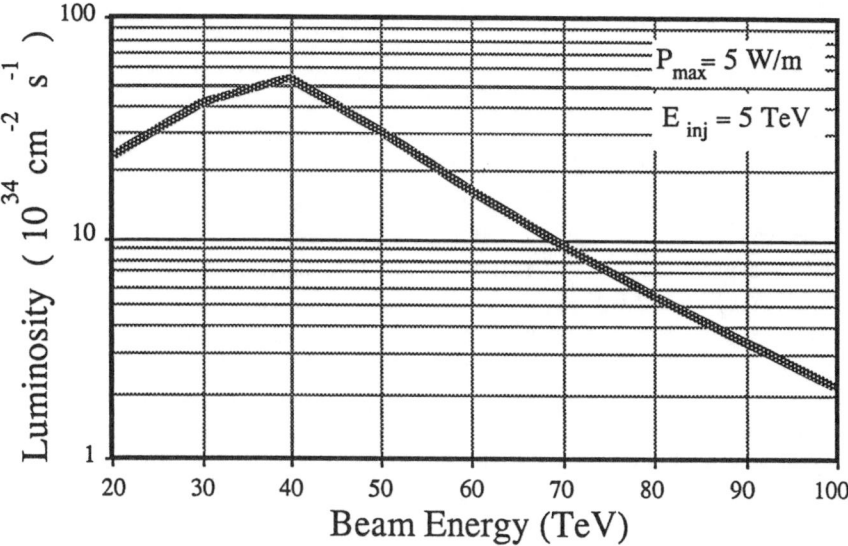

Figure 3. $\mathcal{L}$ versus energy for an ELOSIATRON with 5 W/m of radiation allowed on the walls of the vacuum chamber.

The design code also allows one to search for an optimum value of the dipole field. An examination of the variation of various characteristics of ELN with B_{dipole} for operation at 10^{35} cm^{-2} s^{-1} is shown in Fig. 4. It is difficult to draw any final conclusions about the choice of B_{dipole} with an adequate model of the variation in magnet cost with strength, length and aperture.

If new materials allow one to increase the radiation load to 100 W/m, a 100 TeV/beam, the ELOISATRON could provide a luminosity of $\approx 10^{36}$ cm^{-2}s^{-1}, especially when one operates the collider at intermediate energies (Figure 5). Realizing this latter possibility will depend on the existence of a broadly based experimental research effort prior to the final machine design. Another approach to allowing very large radiation loads it to search for magnet designs that would allow the radiation to be deposited on warm surfaces. In either case and most importantly, the possibility of achieving top luminosity (> 10^{36} cm^{-2} s^{-1}) must be incorporated <u>ab initio</u> into the design, not considered as an after thought.

Table 1. Parameter set for ELN at 10^{34} cm^{-2}s^{-1} with present technology from ELOISASCALE

Ring charcteristics		Interaction points	
* Max Energy (TeV)	100	* Number of IPs	2
* B dipole - max (T)	10	* Beta* (m)	0.5
* Dipole fraction	0.765	* Crossing angle (mr)	0.25
To - rev time (s)	9.13E-04	* L* - long range (m)	170
Circum (km)	273.78	* Crab cross - Y / N	y
		Beta* scale (m)	0.98
Beam characteristics		Sigma-z (cm)	6.0
* Energy (TeV)	100	I* to Q1 - (m)	44.7
* Norm emit@E (mm-mrad)	0.03	Σ-IP (μm)	0.5
* Bunch space buckets	10	Δnu-HO / IP	1.2E-02
gamma	1.1E+05	Δnu-LR	0.0E+00
Bunch space (m)	6.59	Δnu-tot	0.024
Number of bunches	4.2E+04	R (lum correct)	1.00
* NBunch (nC) < 0 for input	0.0	Δt crossing (ns)	22.0
Bunch population	2.9E+09	Interact / crossing	33.6
Current (A)	0.02	T-lum/t-damp-perp	6.9
		Luminosity half-life (hr)	7.5
Injection / Energetics		Luminosity / IP	1.1E+34
* Injection Energy (TeV)	5		
Fill time (hr)	2.0	**Instabilities**	
Stored Energy (GJ)	2.04	* R-pipe (cm)	2
D(E,Q1) MGy/yr	53.1	* Operating temp (°K)	20
Hadrons in Q1 (W/kg)	5.3	* Injection temp (°K)	4
Q1 survival (months)	23	Resistive wall -turns	210.5
Debris (kW per side)	20.0	RW at injection - turns	16.9
		μwave Z/n @ inj (ohms)	165.4
Synchrotron radiation			
* Max P(W/m) on walls	2	**Vacuum**	
Uo (GeV/turn)	2.3E-02	* Op. pressure (nTorr)	1
Damping decrement	2.3E-07	* Desorb coeff.	0.001
E- Damping time (s)	3.9E+03	E-crit (oV)	1.0E+04
Power (W)	5.2E+05	N-gamma (s-1 m-1)	3.6E+15
Power density (W/m)	1.9	Req. pumping (L/s/m)	107.7
Parasitic heat per beam		**RF systems**	
Resitive wall (kW)	0.0	* rf -Frequency (Mhz)	455
RF-HOMs (kW)	0.1	N cavities/ring	293
P to Compressors (MW)	39.5	Max. P to klystrons (MW)	3.4

Table 2. Parameter set from the computer code, ELOISASCALE for the ELN at 10^{35} cm^{-2}s^{-1}.

Ring charcteristics		Interaction points	
* Max Energy (TeV)	100	* Number of IPs	2
* B dipole - max (T)	10	* Beta* (m)	0.5
* Dipole fraction	0.765	* Crossing angle (mr)	0.15
To - rev time (s)	9.13E-04	* L* - long range (m)	170
Circum (km)	273.78	* Crab cross - Y / N	n
		Beta* scale (m)	0.98
Beam characteristics		Sigma-z (cm)	6.0
* Energy (TeV)	100	Σ-IP (μm)	1.7
* Norm emit@E (mm-mrad)	0.32	Δnu-HO / IP	1.0E-02
* Bunch space buckets	9	Δnu-LR	1.6E-04
gamma	1.1E+05	Δnu-tot	**0.021**
Bunch space (m)	5.93	R (lum correct)	0.97
Number of bunches	**4.6E+04**	Δt crossing (ns)	**19.8**
* NBunch (nC) < 0 for input	0.0	Interact / crossing	**271.2**
Bunch population	**2.8E+10**	T-lum/t-damp-perp	8.1
Current (A)	**0.23**	Luminosity half-life (hr)	**8.7**
		Luminosity / IP	**1.0E+35**
Injection / Energetics			
* Injection Energy (TeV)	8	**Instabilities**	
Fill time (hr)	2.0	* R-pipe (cm)	3.5
Stored Energy (GJ)	**21.38**	* Operating temp (°K)	60
D(E,Q1) MGy/yr	**476.0**	* Injection temp (°K)	20
Hadrons in Q1 (W/kg)	47.6	Resistive wall -turns	67.7
Debris (kW per side)	**179.1**	RW at injection - turns	**8.6**
		μwave Z/n @ inj (ohms)	**28.0**
Synchrotron radiation			
* Max P(W/m) on walls	20	**Vacuum**	
Uo (GeV/turn)	2.3E-02	* Op. pressure (nTorr)	5
Damping decrement	2.3E-07	* Desorb coeff.	0.001
E- Damping time (s)	3.9E+03	E-crit (eV)	1.0E+04
Power (W)	5.5E+06	N-gamma (s-1 m-1)	3.8E+16
Power density (W/m)	**20.0**	Req. pumping (L/s/m)	**225.9**
Parasitic heat per beam		**RF systems**	
Resitive wall (kW)	4.5	* rf -Frequency (Mhz)	455
RF-HOMs (kW)	10.3	N cavities/ring	293
P to Compressors (MW)	**119.0**	Max. P to klystrons (MW)	**34.2**

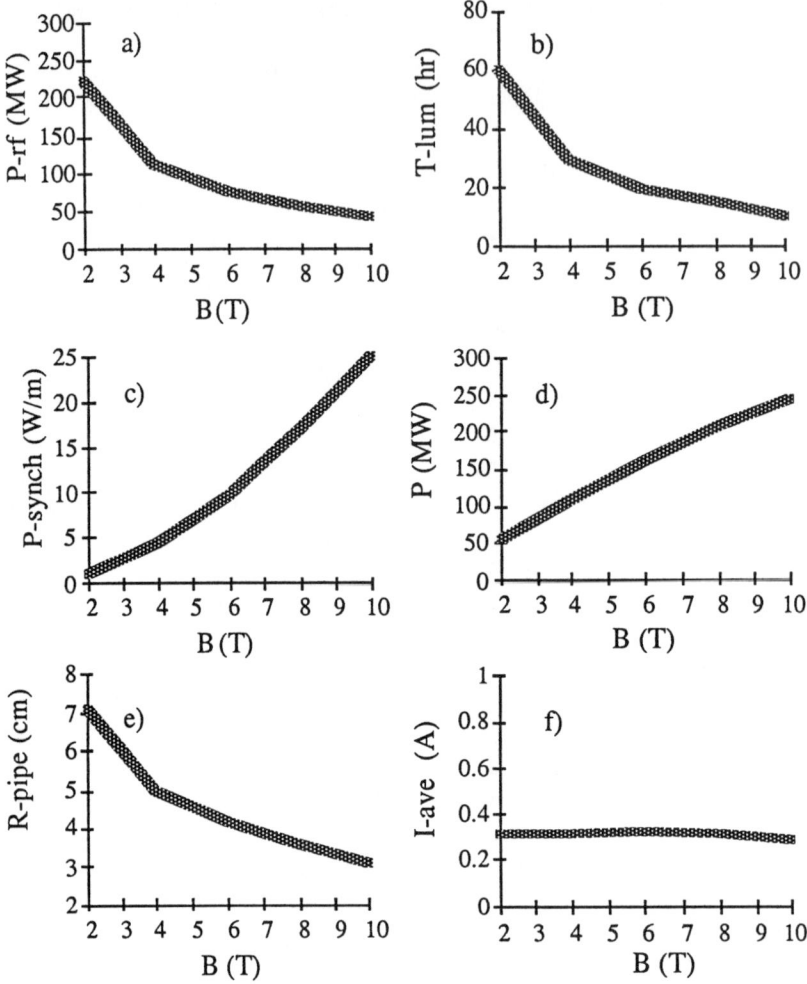

Figure. 4 Variation of ELN charcteristics with dipole field for ELN, a 100 TeV per beam proton supercollider, with a luminosity of 10^{35} cm^{-2} s^{-1}. Panel a) RF-power v. B_{dipole}, Panel b) Luminosity lifetime v. B_{dipole}, Panel c) Thermal load on vacuum chamber walls v. B, Panel d) Total operating power from mains v. B_{dipole}, Panel e) Vertical aperture to control resistive wall modes v. B_{dipole}, and Panel f) Average current v. B_{dipole}

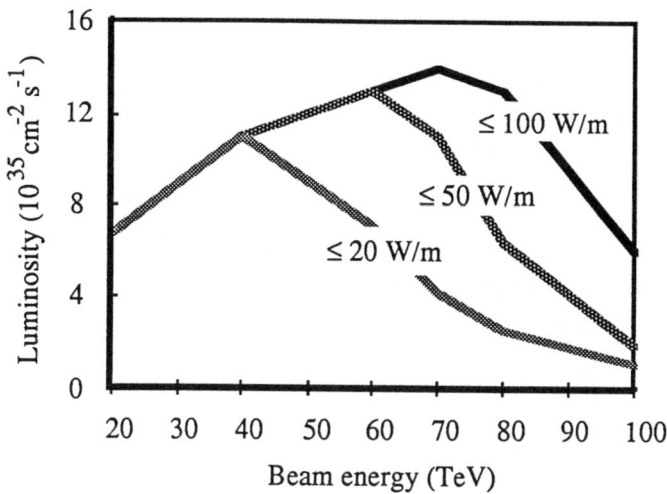

Figure 5. Luminosity limits for ELN if means of handling large radiation loads are developed

With respect to the longest term future, an examination of the systematics of collider design suggests the ultimate potential of conventional storage ring technology in the exploration of the energy frontier of elementary particle physics. If the vacuum chamber of the proton storage ring can be operated at or above room temperatures, then one should be able to construct a hadron collider with a center of mass energy of ≈ 1 PeV and a luminosity $>10^{36}$ cm^{-2} s^{-1} at its peak energy. Such a machine, with a circumference twenty times larger than the SSC (see Table 3) may well be the ultimate hadron supercollider, i.e., the Ultimate ELOISATRON (UELN). This exercise in exploring the limits of conventional technologies also suggests the regime in which linear colliders might be applicable to protons.

Table 3. Characteristics of the Ultimate ELOISATRON

Center of mass energy	1 PeV
Circumference	1600 km
B_{dipole}	20 T
Beam energy	500 TeV
Beam current	200 mA
Mains power	2 GW
$\langle \mathcal{P}_{synch} \rangle$	10 kW/m
Interaction regions (IR)	2
Limiting technology	Magnet survival at IR
Tune shift	0.01 per IR
Luminosity	$\approx 10^{36}$ cm^{-2} s^{-1}

Linear UELN

If the length of each linac of a linear UELN is to no longer than the radius implied by Table 3, then a linac must be able to achieve gradients >> 1 GeV/m. The luminosity is directly related to the average power in the beams by

$$\mathcal{L}(10^{33} \text{ cm}^{-2}\text{s}^{-1}) = \frac{D H_D}{30} \left(\frac{1 \text{ mm}}{\sigma_z}\right) \left(\frac{P_{beam}}{1 \text{ MW}}\right) \quad (3)$$

where H_D is the luminosity degradation due to the pinch effect and where D is the disruption parameter that measures the pinch;

$$D = \frac{r_p N_B \sigma_z}{\gamma \sigma_{x,y}^2} = r_p N_B \left(\frac{\sigma_z}{\beta^* \varepsilon_n}\right). \quad (4)$$

The disruption is related to the tune shift in the ring by

$$D(\text{ring}) = 4\pi \Delta v_{HO} \left(\frac{\sigma_z}{\beta^*}\right). \quad (5)$$

For D < 2, the value of $H_D \approx 1$. At 500 TeV β^* is limited to be ≈ 2 m; the normalized emittance, ε_n, will be difficult to make less than 10^{-7} m-rad. For a 100 GHz accelerating field, the bunch length $\sigma_z \leq 10^{-6}$ m. Thus the quantity in parentheses in Eq. (5) is of order 1 m^{-1}. Even were it possible to generate bunches of 100 nC with such low emittances and to preserve the emittance in the presence of the extremely large wake fields in the linear accelerator, it would be difficult for $r_p N_B$ to exceed 10^{-6} m. Hence the disruption in a proton linear collider will be exceedingly small. Therefore, from (3) one sees that achieving a luminosity of 10^{33} cm^{-2}s^{-1} will require an average power of 30 GW per beam. As D decreases and P_{beam} increases with increasing energy, the practicality of a linear proton collider actually diminishes at even higher energies (above 500 TeV). Therefore, the ultimate hadron supercollider should be a synchrotron.

CONCLUSIONS

On the basis of a systematic parameter search it appears possible to build an ELOISATRON operating at 100 TeV per beam with a luminosity exceeding 10^{34} cm^{-2}s^{-1} by using presently available technology such as that being incorporated in the designs of SSC and LHC. Such a supercollider would have the physics reach of a 10 TeV e$^+$–e$^-$ linear collider, for which no reasonable design concept now exists. Assuming moderate advances in the state of accelerator technology during its design cycle the ELOISATRON could be expected to operate at luminosities $\approx 10^{35}$ cm^{-2}s^{-1} at 100 TeV/beam. Even higher luminosities would obtain at lower energies. With advanced technologies but based upon conventional approaches, a collider, UELN, with an energy five times that of ELOISATRON appears be possible.

ACKNOWLEDGEMENTS

The author wishes to thank Prof. Antonino Zichichi for his encouragement in undertaking this study and for his kind hospitality at the Ettore Majorana Centre for Scientific Culture. Conversations with Robert Siemann (SLAC), Robert Palmer, and Alex Chao have also been extremely helpful in framing the analytical approach. The author's participation in Snowmass 1990 also provide a broad source of the physics embodied in the ELOISASCALE design program. This work was partially performed under the auspices of the Lawrence Livermore Laboratory for the U. S. Dept. of Energy under contract W-7405-eng-48.

POSSIBLE APPLICATIONS OF PLASMA LENS IN HIGH ENERGY PHYSICS[*]

PISIN CHEN
Stanford Linear Accelerator Center
Stanford University, Stanford, CA 94309 USA

ABSTRACT

The concept of the self-focusing plasma lens in various beam-plasma interaction regimes is reviewed. We found that in order for current neutralization to occur, it is only necessary to attain the condition $k_p\sigma_x \gtrsim 1$, and not $k_p\sigma_y \gtrsim 1$, for flat beams. This helps to substantially reduce the required plasma density for beamstrahlung suppression. We also report on a recent calculation on the detector backgrounds induced by a plasma lens. It is shown that these backgrounds are within the tolerance of all major components in a NLC-like detector. Finally, one other potential application of plasma lens for $\gamma\gamma$ colliders is discussed.

1. INTRODUCTION

It is my great honor and privilege to contribute this paper in honor of Andy Sessler's 65th birthday. We all admire his seminal contributions to beam physics which span over four decades in his outstanding career. One of the endless list of his contributions is the understanding and the application of the plasma lens. So I consider it most proper to report on the status and possible applications of the plasma lens in high energy physics as a tribute to Andy.

Although the concept of plasma lens [1] has been extensively studied theoretically [2-11], and has been reasonably verified in several low energy, low density experiments [12-14], the utility of the plasma lens in high energy physics has yet to be proven. It is necessary to further demonstrate the plasma focusing of high energy, high density electron and *positron* beams in the parameter regime and the physical setting close to the true future high energy linear colliders. But more importantly, one needs to demonstrate, both theoretically and experimentally, that the effects of background events due to plasma lenses located deep inside a high energy collider detector are within detector tolerance.

In this paper we first review the concepts and mechanisms of the self-focusing plasma lens in different beam-plasma interaction regimes, with attention focused on a new realization on the current neutralization, which helps to reduce the required plasma density for beamstrahlung suppression. Next we discuss a recent calculation on the plasma lens induced backgrounds [15], and show that these backgrounds are within the detector tolerance in a NLC-like detector. We then review the existing low energy experiments and a proposal to perform a series of plasma lens experiments at the Final Focus Test Beam (FFTB) at SLAC [16,17]. Application of plasma lens other than enhancing the luminosity and suppressing the beamstrahlung in e^+e^- colliders, such as the *overfocusing* of the spent electron beams in $\gamma\gamma$ colliders [18], is discussed.

---

[*] Work supported by Department of Energy contract DE–AC03–76SF00515.

2. DIFFERENT REGIMES OF PLASMA LENS

A. THE SELF-FOCUSING REGIME

When $k_p\sigma \ll 1$, where k_p is the plasma wavenumber, $k_p = \sqrt{4\pi r_e n_p}$ (r_e is the classical electron radius and n_p is the plasma density) and σ is the transverse beam size, the plasma response to the incoming beam is such that it quickly neutralizes the space charge of the beam, yet the return current runs mostly outside the beam volume. This leaves the magnetic self-force in the beam unbalanced, and the self-focusing occurs. When the ambient plasma density n_p is greater than the density of the incoming charge particle beam, n_b, one further refers the situation as the *overdense regime* of beam-plasma interaction. One of the main features of the self-focusing plasma lens in this regime is that it applies equally to both electron and positron beams [1]. In the overdense regime the plasma is only mildly perturbed by the intruding particle beam. With enough reserve, the plasma electrons will either enhance or deplete its density against the less mobile ions inside the beam volume, depending on the sign of charges of the beam, so as to neutralize the excessive space charge introduced by the impinging beams. This leaves the azimuthal magnetic field of the beam unbalanced and a *self-focusing* force is triggered. The focusing force thus induced is a direct reflection of the local beam density, and therefore is nonlinear around the edge of the beam. Moreover, as the particle distribution necessarily varies along the direction of beam propagation, the focusing strength also varies from the head to the tail of the beam.

Let the spatial distribution of a round particle beam be

$$\sigma(r,z) = n_b f(r) g(\zeta) \quad , \tag{1}$$

where $\zeta \equiv z - ct$. n_b is the reference beam density near the center of the bunch, and $n_b = N/(2\pi)^{3/2}\sigma_z\sigma_r^2$ for bi-Gaussian beams. When the effect of the returned current is ignored, it can be shown [1] that the plasma wakefield perturbed by such a beam induces a self-focusing effect, where the focusing strength, K, for both electron and positron beams can be expressed in terms of the Green's functions for the wakefield:

$$K(r,\zeta) = \frac{4\pi r_e n_b}{\gamma k_p^2} \frac{\partial^2 F(r)}{\partial r^2} G(\zeta) \quad , \tag{2}$$

where

$$F(r) = k_p^2 \Big[\int_0^r r' dr' f(r') I_0(k_p r') K_0(k_p r) + \int_r^\infty r' dr' f(r') I_0(k_p r) K_0(k_p r') \Big] \quad ;$$

$$G(\zeta) = k_p \int_\zeta^\infty d\zeta' g(\zeta') \sin k_p(\zeta' - \zeta) \quad . \tag{3}$$

I_0 and K_0 are the modified Bessel functions. It is clear that K closely reflects the distribution of the beam. Indeed, for a beam length much longer than the reduced plasma wavelength, or k_p^{-1}, $G(\zeta)$ closely follows $g(\zeta)$, the longitudinal beam distribution. The above expression clearly suggests that such a focusing is aberration-prone.

Triggered by these considerations, an opposite regime of plasma lens, i.e., the underdense regime where $n_p < n_b$, was proposed [5,6]. In this regime the plasma perturbation is so strong that the plasma electrons are quickly rarefied from the beam volume (so-called *wave-breaking*), leaving the less mobile ions to focus the electron beam. As for the positron beam, the mechanism is entirely different. Here the plasma electrons are drawn in from outside the beam volume and oscillate across the beam cross section. During the time when these electrons stay inside the beam volume there will be an excessive negative space charge, which in turn attracts the positron inward, thus the self-focusing. Although the focusing forces in the underdense regime are generally not equal and are also weaker than that in the overdense regime, theoretical calculations and simulations show that they are indeed much more linear transversely and more uniform longitudinally.

However, the real purpose of a final focus is to attain high luminosity, which is the ultimate goal of every high energy collider. In the case of linear collider, the *disruption effect* [19] between the colliding e^+e^- beams is a very nonlinear function of the beam current density. Simulations on the effective luminosities due to plasma focusing over a range of n_b/n_p (from underdense to overdense) [9] have shown that, albeit the nonlinearities and the longitudinal variations, the aberration-prone overdense regime, with its stronger focusing available, still results in a higher luminosity than the underdense regime, due to a stronger head-start that triggers a stronger beam-beam disruption [9]. Nevertheless, from the backgrounds considerations there is still an advantage for the underdense lens.

B. THE FIELD-COMPENSATION REGIME

When $k_p\sigma \gtrsim 1$, the plasma return current will penetrate into the beam volume. In this situation the azimuthal magnetic field in the beam is also shielded. The self-focusing force is then diminished in this regime. Earlier this effect was observed in a computer simulation and was suggested as a means to suppress the undesirable *beamstrahlung* during beam-beam interaction in linear colliders [2]: "Having a neutral plasma at the interaction point necessarily modifies the beam-beam interaction...The fact that the plasma partially neutralizes the currents reduces the disruption effect...the collective field strengths are reduced, and the beamstrahlung effect is also supposed to be weakened." This suggestion was largely forgotten until the discovery of the *coherent pair creation* backgrounds [20] in linear collider beam-beam interaction, when people became very worry about the large quantities of deleterious beamstrahlung-induced backgrounds, and therefore the future of linear colliders. Without noticing the above mentioned earlier suggestion, Andy and his collaborators reinvented the idea of beamstrahlung suppression in the far-overdense beam-plasma interaction regime [7]. Their detail work shows the amount of suppression achievable for a chosen beam-plasma condition.

In the flat beam geometry, i.e., for beams with aspect ratio $R = \sigma_x/\sigma_y \gg 1$, Whittum et at. [7] invoked the following approximation: $\partial/\partial x \ll \partial/\partial y, B_z \approx 0, B_y \ll B_x$, and $E_x \ll E_y$, and derived an equation which governs B_x:

$$\frac{\partial^2 B_x}{\partial y^2}(y,\tau) = \frac{4\pi}{c}\frac{\partial J_{bz}}{\partial y}(y,\tau) + k_p^2 \int_{-\infty}^{\tau} d\tau' \frac{\partial B_x}{\partial \tau}(y,\tau') \exp\left\{ -\int_{\tau}^{\tau} d\tau'' \nu(\tau'') \right\} , \quad (4)$$

where $\tau = t - z/v$ is the beam-plasma interaction time, J_b is the beam current, and ν is the collision rate. In the collisionless regime, the solution was found to be [7]

$$B_x(y,\tau) = \frac{2\pi}{\omega_p} \int_{-\infty}^{\infty} dy' \exp(-k_p|y-y'|) \frac{\partial J_{bz}}{\partial y'}(y',\tau) \quad . \tag{5}$$

It is clear that, as the dominant dimension involved in this calculation is the minor dimension, y, the relevant quantity in the consideration is σ_y relative to the plasma wavelength, or $k_p \sigma_y$. For example, for $k_p \sigma_y \sim 1$, B_x is reduced by roughly fivefold.

One common concern regarding plasma lenses has been the backgrounds induced by the beam-plasma interaction. For beamstrahlung compensation the required plasma density is even higher. In the above example, and for the case of the SLAC design of the Next Linear Collider (NLC) where $\sigma_y \sim 3$nm, we need a plasma of density $n_p \sim 3 \times 10^{24} \mathrm{cm}^{-3}$! This extremely high density may even be physically impossible to attain, let alone the backgrounds concerns.

However, a closer look at the above derivation soon reveals that the assumption, $B_y \ll B_x$, is not universally true everywhere inside the beam even for a very flat beam. We know that the magnetic field of a finite size beam always form close contours, which cut across the beam cross section. Indeed, using the Bassetti-Erskine formula, one can derive the locations of the maxima of B_y along the x-axis and B_x along the y-axis [21], respectively:

$$\begin{aligned} x_m/\sigma_x &= 1.307 + 0.393/R^2 \quad ; \\ y_m/\sigma_y &= \sqrt{2\ell n R} + \sqrt{\pi/2R^2} \quad . \end{aligned} \tag{6}$$

And the corresponding maximum field-strengths are:

$$\begin{aligned} B_{ym} &\simeq (r_e mc^2 N/\sigma_x \sigma_z) \exp(-z^2/2\sigma_z^2) \sqrt{2/\pi} \left[1 - 0.726/R\right]/1.307 \quad , \\ B_{xm} &\simeq (r_e mc^2 N/\sigma_x \sigma_z) \exp(-z^2/2\sigma_z^2) \left[1 - \sqrt{\pi \ell n R/4R^2}\right] \quad . \end{aligned} \tag{7}$$

We note that $B_{ym}/B_{xm} \to 0.610$ as $R \to \infty$. So there is always a vertical component of the magnetic field with a strength comparable to that of the horizontal one.

For the purpose of stressing the point, bearing in mind, however, that the issue at stake is actually a two-dimensional one (and therefore the x and y dimensions should really be coupled), we can express the resultant B_y due to the return current by exchanging the roles of x and y in Eq.(5):

$$B_y(x,\tau) = \frac{2\pi}{\omega_p} \int_{-\infty}^{\infty} dx' \exp(-k_p|x-x'|) \frac{\partial J_{bz}}{\partial x'}(x',\tau) \quad . \tag{8}$$

It is very interesting to see that in this notion the relevant parameter for plasma beamstrahlung suppression is $k_p \sigma_x$ rather than $k_p \sigma_y$. As $\sigma_x/\sigma_y \sim R \gg 1$, and since $k_p = \sqrt{4\pi r_e n_p}$, to achieve the same amount of suppression, one would need a plasma density a factor $1/R^2$ smaller than the previous value. For NLC, this corresponds to a 10^{-4} reduction (or $n_p \sim 3 \times 10^{20} \mathrm{cm}^{-3}$) in plasma density.

Intuitively, this somewhat surprising revelation can be easily appreciated. While B_{xm} is comparable to B_{ym}, the former decays as $1/x$ outside σ_x, while the later maintains its strength many σ_y's beyond the beam boundary. So it is much easier for the return current to penetrate the beam horizontally along its major axis.

Three comments are in order. In the above discussion the return current in x and y dimensions are treated independently. In this approach even if we manage to suppress the B_y, there is still an unsuppressed B_x since $k_p\sigma_y \ll 1$ if we choose the condition $k_p\sigma_x \sim 1$. Nevertheless it is possible that once the return current penetrates into the beam along the major axis (i.e., x-axis), due to the constraint from the continuity of magnetic field contours, B_x will likely be suppressed as well [22]. This needs to be further investigated. Secondly, it is worth to mention that in the effort to suppress beamstrahlung photon yields, both the *interaction cross section* and the *luminosity* play important roles. Namely, even if B_x is not entirely suppressed near the center of a flat beam (therefore the probability for beamstrahlung in that area is still high), as long as a large fraction of the beam transverse area is shielded with return current (therefore the beamstrahlung luminosity becomes low), the average yield of beamstrahlung photons will be largely suppressed. Finally, another approximation was made in the derivation in Ref. [7]. Namely, the plasma was assumed to be unmagnetized, i.e., the $\vec{v} \times \vec{B}$ force was neglected. In fact, with the NLC parameters the corresponding cyclotron frequency, $\omega_H = eH/mc$, is much larger than the $1/\tau_r$, where τ_r is the beam current rise time. So its effect on the return current may not be negligible [22].

It is clear that more studies are needed before we fully understand the nature of current neutralization, especially in the flat beam regime.

3. THICK LENS AND ADIABATIC FOCUSING

The focusing of particle beams by strong lenses is accompanied by radiation due to the transverse acceleration. The radiation is statistical in nature and begins to degrade the achievable focusing when the lenses are strong. For discrete lenses there is consequently a limit to the attainable spot size which is called the synchrotron radiation limit or the Oide limit [23]. This limitation occurs because of the chromatic aberration developed through the drift space between the focusing lens and the focal point. While one cannot turn off the radiation, one can indeed in principle eliminate the drift space and conceive a thick, continuous lens, and maybe the Oide limit can thus be evaded. Indeed, it was my great privilege to have collaborated with Andy (together with Oide and Yu) [10] to propose the concept of a plasma-based *adiabatic lens*, where the focusing force is adiabatically tapered, as a means to overcome the Oide limit. We found that in an adiabatic focuser the Oide limit on the attainable beam size is replaced by the following formula:

$$\sigma \gg 1.39 \times 10^{-8} \left(\frac{\epsilon_n}{\epsilon_c}\right)^2 \exp(-1.12\epsilon_c/\epsilon_n)\text{m} \quad, \tag{9}$$

where

$$\epsilon_c \equiv \frac{3^{3/2} \cdot 15^3}{2^3 \cdot 4^2 \cdot 22} \frac{\lambda_c}{\alpha^3} = 6.17 \times 10^{-6}\text{m} \quad, \tag{10}$$

and λ_c is the Compton wavelength and α is the fine structure constant. With the choice of the initial Twiss parameter, or the *adiabaticity*, $\alpha_0 = \sqrt{3}$ and $\epsilon_n = \epsilon_c/10$, we find that $\sigma \gg 2.7 \times 10^{-9}$m, which is indeed smaller than the Oide limit in this case.

4. DETECTOR BACKGROUNDS FROM PLASMA LENS

Plasma lens induced backgrounds in the SLD (Stanford Linear Colliders Detector) has been studied earlier [24]. Further detailed calculations based on the NLC parameters and a NLC-like detector have been preformed recently [15]. Here we briefly review the essence of Ref. [15]. The sources of backgrounds for particle detectors from plasma lenses can be divided into three kinds, namely, electrons/positrons, protons and photons. They originate from different elementary physical processes underlying the interactions of the incoming electron or positron beams with the plasma near the interaction point. The electrons and positrons come from the scattering of the e^+ or e^- beam with the electrons of the plasma. The processes for producing electrons and positrons are:

Bhabha scattering: $e^+e^- \to e^+e^-$,
Møller scattering: $e^-e^- \to e^-e^-$,
Elastic scattering: $e\,p \to e\,p$,
Inelastic scattering: $e\,p \to e\,X$.

The hadronic backgrounds come from the elastic and inelastic scatterings of the e^+ or e^- beam with the protons in the plasma. The processes for producing hadrons are:

Elastic scattering: $e\,p \to e\,p$, $\quad \gamma\,p \to \gamma\,p$,
Inelastic scattering: $e\,p \to e\,X$, $\quad \gamma\,p \to X$.

There are three mechanisms for producing the incident photons, which are then scattered by the plasma electron into large angles: synchrotron radiation at the final focus quadrupole, synchrotron radiation as a consequence of plasma focusing, and bremsstrahlung of the e^+ or e^- beam in the plasma. The underlying physical process for producing photons into the detector is the Compton scattering off electrons and protons in the plasma.

Compton scattering: $\gamma\,e \to \gamma\,e$, $\quad \gamma\,p \to \gamma\,p$.

For our calculations, we have taken NLC beam energy E_0 at 250 GeV, $N = 0.65 \times 10^{10}$, $\sigma_x = 300$ nm, $\sigma_y = 3$ nm, $\sigma_z = 100$ μm, the hydrogen plasma density $n_p = 10^{18}$ cm^{-3} and $L = 2$mm. Hence $\mathcal{L}=6.5 \times 10^{26}$ cm^{-2} per bunch crossing. For a train of $n_b = 90$ bunches, $\mathcal{L}=5.85 \times 10^{28}$ cm^{-2} for a bunch train. The time structure of NLC is that the RF repetition rate is 180 Hz, and within each filling there are $n_b = 90$ bunches, with a bunch to bunch separation of 1.4 nsec. To be conservative, we assume that the detector cannot resolute the backgrounds down to nsec level. So we compute the yield per bunch train. Note that the numbers thus obtained are really very conservative, considering the capability of fast response in modern detectors [25].

We further note that the detector for NLC will not occupy all 4π solid angle, due at least to the masking angle, $\theta_m = 150$mrad, around the entrance and exit beam pipes. Since for leptonic and hadronic events the center-of-mass frames are highly forwardly boosted, the cross sections back in the Lab frame are highly forward peaked. Because of the masking, the resultant partial cross section is minimal. However, the incoming photons are typically much softer and they contribute as the primary source of plasma lens induced backgrounds.

There are three major components in a high energy particle detector: the vertex detector, the drift chamber, and the calorimeter. As discussed above, the leptons and hadrons scattered into the fiducial volume of the detector can be neglected. Thus we expect the presence of a plasma lens near the interaction point will not pose problems to the calorimeter. The photon sources will, however, create backgrounds for the vertex detector and drift chamber. A brief description of the general characteristics of a vertex detector for the NLC can be found in Ref. [26]. In our case the plasma scattered photons cover a wide range of energy. Different components of the future linear colliders are sensitive to certain energy windows of the energy spectrum of the outcoming photons. For the vertex detectors, the serious range for background photons is between 4 keV and 100 keV. When imposing these energy cuts, the number of background photons for the vertex detector is reduced to 3600. For a vertex detector with 2.5 cm radius and with a length covering angles greater than 150 mrad ($\cos\theta \leq 0.99$), the density of photon tracks per unit surface area is $\sim 0.14/\text{mm}^2$, which is only about 15% of the tolerable level of background, which is $\sim 1/\text{mm}^2$. Note that the choice of $\cos\theta \leq 0.99$ for the vertex detector is very generous. The backgrounds will be further reduced if, for example, $\cos\theta \leq 0.90$. The next layer in the detector may well be the drift chamber which can tolerate about 10,000 incident photons, of which typically no more than of the order of 100 convert. Only photons of more than, say 100 keV, need to be considered; the conversion products of those of less than $\sim$ 100keV will not form track segments inside the drift chamber for a typical magnetic field. The total number of background photons is about 4,700, which is, again, below the acceptable limit.

The partial cross sections and the detector responses to the various sources of plasma lens induced backgrounds are summarized in the following Table. We emphasize again that the yields are calculated per train of 90 bunches, with a total time span of $\sim$125 nsec and the time separation from the next bunch train of $\sim$ 5 msec. This should be very conservative.

Background Source	Partial Cross Section $(\text{cm}^{-2})(\cos\theta \leq 0.99)$	Vertex Detector	Drift Chamber
Bhabha and Møller	0	0	0
Elastic $ep : e$	0.103×10^{-45}	negligible	negligible
p	0.613×10^{-39}	negligible	negligible
Inelastic $ep : e$	0.132×10^{-33}	negligible	negligible
charged hadrons	0.396×10^{-29}	0.23	0.23
Inelastic γp: charged hadrons	0.313×10^{-28}	1.8	1.8
Compton from quadrupole	0.995×10^{-25}	2000 γ's	2600 γ's
Compton from plasma focusing	0.548×10^{-25}	990 γ's	1800 γ's
Compton from bremsstrahlung	0.119×10^{-24}	540 γ's	270 γ's

Table 1: Summary of background sources from a plasma lens in NLC for a train of 90 bunches. Each bunch has 0.65×10^{10} particles, and the plasma density is 10^{18} cm^{-2}.

5. EXPERIMENTAL PROGRESS

There are several low-energy, low-density beam experimental results which confirm the theory of the beam-plasma interaction. In the Argonne experiment, [12] a 21 MeV electron beam with $2.5 - 4 \times 10^{10}$ cm^{-3} particles per bunch was sent through a thick plasma of 35 cm long followed by a 15 MeV low intensity witness beam, with adjustable delay and transverse offset. The plasma was created in a hollow cathode arc and had a density of $0.7 - 7 \times 10^{13}$ cm^{-3}. The size of the beam was seen to decrease from $\sigma = 1.4$ mm without plasma to $\sigma = 0.91$ mm with plasma, roughly the predicted equilibrium size in a long plasma. Before reaching this equilibrium size, the beam was expected to have been as much as 3 times smaller, but unfortunately there was no measurement of the beam size at intermediate positions in the plasma.

In the Tokyo University experiment a round 18 MeV electron beam was sent through a thin plasma lens and focusing was observed. [13] They measured the beam size at three phosphor screens along a drift downstream from the plasma chamber. The beam density was about 1.2×10^{10} cm^{-3} and the plasma density was increased up to about 1.4×10^{11} cm^{-3}, created before the passage of the electron bunch by a current discharge. This experiment confirms the theory of thin plasma lens.

The most decisive experiment came from the recent effort by the UCLA group [14], where the *dynamic focusing* of a 3.8 MeV electron bunch, a few collisionless skin-depth long $\sim 3c/\omega_p$, by an overdense, thick plasma lens has been demonstrated. Because of the longitudinal variation as discussed in Section 2, the head of the bunch is virtually unaffected by the lens while the rest is focused to varying degrees. Time-resolved measurements performed 31 cm downstream of the plasma lens show that, in time, the bunch pinches from an initial size of 2.7 mm (FWHM) to about 0.57 mm and then expands, in reasonable agreement with theory.

While these experimental results have been useful, the experience is insufficient to design or evaluate a plasma lens in a high energy collider detector. Systematic evaluations of methods and experimental techniques with realistic high energy beams are now required. It is necessary not only to understand the fundamental physics of plasma focusing, but also to make the technological and practical advances so that plasma focusing devices may be used in future accelerators. The nominal colliding beam density at the SLC is about 1×10^{18} cm^{-3} and in the next generation linear colliders at 0.5 TeV center-of-mass energy it is above 1×10^{19} cm^{-3}. These densities are 6 to 7 orders of magnitude higher than those in the Argonne, Tokyo, and UCLA experiments.

A consortium of more than 30 physicists from 15 institutions, with Andy's group from LBL included, is now proposing [16] a series of plasma lens experiments (E150) at SLAC's Final Focus Test Beam (FFTB), which was designed for experimentations with the final-focusing system and the diagnosis of beam conditions similar to that in the next generation of linear colliders. The main emphasis of this experiment is not only the demonstration of the self-focusing effect, although that for the positron beams will be the first of its kind. Rather, it is to demonstrate the utility of the plasma lens in high energy physics. From the basic theory [1], the self-focusing effect should behave pretty much the same as long as the beam is relativistic.

The challenge lies mostly in the stringent constraints of a high energy detector. The plasma lens induced backgrounds is certainly one of them, which was addressed in the previous section. Another challenge is the demand for high vacuum along the beam pipe, $\lesssim 10^{-6}$ Torr, while the plasma density commensurate with the NLC beam parameters is $n_p \sim 40$ Torrs. Since it is undesirable to let the high energy beams pass a solid material (for the purpose of emittance preservation, at the least), it is not possible to confine the plasmas in a container. As such the supply and the extraction of the gas/plasma poses a major challenge. Our current approach is a supersonic gas jet (to be ionized by a laser) with minimal cross section and mass flow so as to reduce backgrounds and gas-pumping load. Initial R&D of this approach has been reasonably successful [17]. Together with the recent backgrounds calculations [15], which suggest that the plasma lens induced detector backgrounds should be reasonably minor, The prospect of a real application of the plasma lens is quite promising.

6. APPLICATION IN $\gamma\gamma$ COLLIDERS

So far our discussion has been on the applications of the plasma lens in e^+e^- linear colliders. A different type of linear collider has been proposed [27], where high energy electron beams are converted into photon beams through the Compton scattering against laser beams, and then collide. The scientific motivations and the technical challenges of colliding high energy photon beams has been well-documented [28]. One main technical challenge is the handling of the spent electron beams after the Compton conversion and before the $\gamma\gamma$ collision. The original suggestion [26] was to apply an external bending magnetic field between the conversion point (CP) and the interaction point (IP), so as to deflect the spent electron beam away from the point where photon beams collide. Limited by the available field strength and physical space, this approach imposes a constraint on the attainable $\gamma\gamma$ luminosity [20], due to the need to suppress the coherent pair creation induced by the residual field of the near-by spent beams. Another suggestion [29] is to *do nothing*, so to speak, but just let the spent beams comove with the photon beams and collide. The nature of the e^-e^- beam-beam disruption will make the beams rapidly disperse each other. This approach may indeed work for machines like the Russian VLEPP project, where the vertical disruption is designed to be extremely large. With much milder disruptions such as the NLC, it is unclear if the anti-pinch between the electron beams is strong enough to sufficiently damp the e^-e^- luminosity. Even if it does, since we know for flat beams the beam-fields stretch many σ_y's beyond the beam volume, the residual beam field will still be very intense.

A third approach was recently suggested [30], where the spent electron beam is *overfocused* by a thin plasma lens while the photon beam is inert to it. As a result, the electron beam will be much larger in size when arrives together with the tighter photon beam at the interaction point, and therefore the e^-e^- luminosity is much reduced. But more importantly, the photons, now situate near the center of the electron beams, will see a much reduced beam field and would thus create less backgrounds. Calculations showed that a factor 5 in luminosity can in principle be gained for the case of a $\gamma\gamma$ collider based on NLC parameters if a bend field of 30 kG long is replaced by a plasma lens with density 10^{18}cm^{-3} and 1 mm thick.

Like the other two approaches, there is not without any short-coming in this one. One worry is that the residual gas from the plasma lens (e.g., the gas jet) may spread to the conversion point so that the quality of the converted photon beam may have to be compromised [31]. The best way to address this issue is through experiment. Our initial study of the gas jet indicates that the supersonic jet profile is well-collimated and localized. In addition, the demand for cleanliness at the conversion point coincides with the demand for high vacuum discussed earlier, which is part of the task of the proposed plasma lens experiments at the FFTB.

7. SUMMARY

We have reported in this paper several progress towards the eventual application of plasma lens in high energy physics. Although the physics of plasma lens is reasonably understood, we point out that there is still some room to be improved in field-compensation for flat beams. On the practical side, we showed that all the main components of the detector should survive from the plasma lens induced backgrounds. Thus the implementation of a plasma lens for luminosity enhancement in high energy e^+e^- collisions is feasible without hampering the normal performance of the detector. When applying to $\gamma\gamma$ colliders, we argued that substantial increase in luminosity can be attained using a plasma lens to overfocus the spent electron beam.

Albeit the promise of luminosity enhancement and field compensation, the backgrounds to the detector and the general demand for a super-clean accelerator environment have been the continued worry about the plasma lens among high energy physicists. It is hoped that many critical issues regarding the applicability of plasma lens in high energy physics will be address by the E150 experiment in the coming couple years. In this regard, I am looking forward to many more years of fruitful collaborations with you, Andy. Happy Birthday!

REFERENCES

[1] P. Chen, *Part. Accel.* **20**, 171 (1987).

[2] P. Chen, J. J. Su, T. Katsouleas, S. Wilks, and J. M. Dawson, *IEEE Trans. Plasma Sci.* **15**, 218 (1987).

[3] J. B. Rosenzweig, B. Cole, D. J. Larson, and D. B. Cline, *Part. Accel.* **24**, 11 (1988).

[4] J. B. Rosenzweig and P. Chen, *Phys. Rev. D* **39**, 2039 (1989).

[5] P. Chen, S. Rajagopalan, and J. B. Rosenzweig, *Phys. Rev. D* **40**, 923 (1989).

[6] J. J. Su, T. Katsouleas, J. M. Dawson, and R. Fedele, *Phys. Rev. A* **41**, 3321 (1990).

[7] D. H. Whittum, A. M. Sessler, J. J. Stewart, and S. S. Yu, *Part. Accel.* **34**, 89 (1990).

[8] P. Chen, *Phys. Rev. A* **45**, R3398 (1992).

[9] P. Chen, C. K. Ng, and S. Rajagopalan, *Phys. Rev. E* **48**, 3022 (1993).

[10] P. Chen, K. Oide, A. Sessler, and S. Yu, *Phys. Rev. Lett.* **64**, 1231 (1990).

[11] A. T. Amatuni, S. S. Elbakian, and E. V. Sekhposian (Yerevan Phys. Inst.), LBL-34836, 1993.

[12] J. B. Rosenzweig et al., *Phys. Fluids B* **2**, 1376 (1990).

[13] H. Nakanishi et al., *Phys. Rev. Lett.* **66**, 1870 (1991).

[14] G. Hairapetian et al., *Phys. Rev. Lett.* **72**, 2403 (1994).

[15] P. Chen, C. Ng, and A. Weidemann, "High Energy Detector Backgrounds from Plasma Lenses", 1995; to be submitted to *Nucl. Instr. Meth.*

[16] W. Barletta, et al., *Plasma Lens Experiments at the Final Focus Test Beam*, SLAC-Proposal E150, 1993.

[17] P. Kwok, et al., "Progress on the Plasma Lens Experiments at the FFTB", to appear in the *Proc. Part. Accel. Conf.*, Dallas, Tx, 1995.

[18] S. Rajagopalan, D. B. Cline, and P. Chen, *Nucl. Instr. Meth.* A **355**, 169 (1995).

[19] R. Hollebeek, *Nucl. Instr. Meth.* A **184**, 333 (1981); P. Chen and K. Yokoya, *Phys. Rev.* D **38**, 987 (1988).

[20] P. Chen and V. M. Telnov, *Phys. Rev. Lett.* **63**, 1796 (1989).

[21] P. Chen, "Plasma Focusing and Diagnosis of High Energy Particle Beams," in *Nonlinear and Relativistic Effects in Plasmas*, ed. V. Stefan, AIP Research Trends in Physics, p.219 (1992).

[22] P. Chen and G. Stupakov, in preparation, 1995.

[23] K. Oide, *Phys. Rev. Lett.* **61**, 1713 (1988).

[24] C. Baltay, "Backgrounds in SLD due to the Proposed SLC Plasma Lens", SLD Internal Notes, 1992, unpublished.

[25] P. Chen, T. L. Barklow, and M. E. Peskin, *Phys. Rev.* D **49**, 3209 (1994).

[26] C. J. S. Damerell, "Vertex Detectors at Linear Colliders (Present and Future)", RAL Report RAL-94-096, 1994.

[27] I. Ginzburg, G. Kotkin, V. Serbo, and V. Telnov, *Nucl. Instr. Meth.* **205**, 47 (1983).

[28] *Proc. of Workshop on Gamma-Gamma Colliders*, ed. S. Chattopadhyay and A. M. Sessler, *Nucl. Instr. Meth.* A **355**, 1-194 (1995).

[29] V. Telnov, *ibid.*, p. 3; V. E. Balakin and A. A. Sery, *ibid.*, p. 157.

[30] S. Rajagopalan, D. B. Cline, and P. Chen, *ibid.*, p. 169.

[31] V. Telnov, private communications, April, 1995.

MURA DAYS

Keith R. Symon
University of Wisconsin-Madison, Madison, WI 53706

ABSTRACT

This paper discusses the technical developments in which the MURA group participated, as recalled at this time by the author. Andrew Sessler contributed to many of these developments. Early work at MURA included the development of fixed-field alternating-gradient (FFAG) accelerators, the radial sector FFAG model, the spiral sector model, the development of the theory of rf acceleration in circular accelerators including beam stacking, and the development of colliding beams as a practical experimental goal. Theoretical studies also included nonlinear orbit computations, space charge limits, and analysis of the negative mass and other instabilities. Experimental studies of many of these phenomena used the FFAG models. The cascade synchrotron, developed during the 1959 MURA Summer Study, was the basis for most modern accelerators.

INTRODUCTION

I am honored to be asked to speak at this symposium on the occasion of Andrew Sessler's sixty-fifth birthday. Andy is a valued colleague and a good personal friend. I first met Andy when he joined the MURA group in 1955.

In this talk, I will give a review of the MURA technical program, with particular emphasis on MURA's impact on the field of accelerator technology. Andy played a leading role in many of the developments that I will discuss. I am relying almost solely on my present memory of those events now long past. Therefore you must flavor what I say with the necessary grains of salt, except where it is documented by the figures that I will show you.

FFAG

I conceived the idea of the fixed field alternating gradient accelerator in the Summer of 1954. The same idea was conceived independently by others. In particular, L. H. Thomas had proposed the Thomas cyclotron a number of years earlier, which uses essentially the same idea. We were unaware of Thomas's paper at that time, nor were we aware that much classified work had been done at Berkeley on the Thomas cyclotron. The idea also occurred to Tihiro Ohkawa in Japan at about the same time. There are a number of advantages to using a fixed magnetic field, as in a cyclotron, though they seemed more important at the time than they later turned out to be. By combining a fixed field with alternating gradients, one can design an accelerator which can confine simultaneously particles with a wide range of energies from the injection energy to the output energy. The original idea was to build the magnets in sectors in which the field pattern extends radially outward from the injection radius to the output radius. This was called a *radial sector* accelerator.

The idea intrigued the working group, and in January, 1955, we began the design of the Michigan model, a 400 keV radial sector FFAG electron accelerator, which was to be a proof of principle. It was designed by the group and built in Ann Arbor by Kent Terwilliger and Larry Jones, using parts built at various universities. The magnets were constructed at Purdue, under the direction of Robert Haxby, and brought to Ann Arbor by train by Donald Kerst. Figure 1 shows a photo of the

© 1996 American Institute of Physics

completed machine. You can see the alternating wide and narrow radial sector magnets. The field is reversed in the narrow magnets; hence the field gradient also reverses, so the gradient alternates from one magnet to the next. Of course the orbit is concave outward in the reversed field narrow magnets, so the mean orbit radius is larger by a large factor than if the field were constant all the way around. This is called the *circumference factor* and is about five.

I remember one incident during the design when the group were divided over the question whether the magnets should scale. As noted earlier, the magnetic field pattern extends out radially, with all dimensions increasing in proportion to the radius. This is called (geometrical) scaling. The question was whether the magnets themselves should scale geometrically in this way, which would mean that the magnet apertures would have to increase in proportion to the radius. The magnetic field could be made to scale without requiring the magnets to scale geometrically, but making them scale makes it easier to guarantee that the field pattern scales.

Fig. 1. The Michigan Model

Hence the theorists favored scaling. However, the field strength increases as a power of the radius. Scaling the aperture would have made it harder to make the field increase, so the experimentalists probably favored not scaling the magnets. As in all debates like this, where everyone agreed it could be done either way, we had difficulty arriving at a decision. Don Kerst, our leader, finally put it to a vote. The vote was in favor of scaling the magnets.

First beam in the Michigan model was achieved in March, 1956. Although a high energy FFAG accelerator has never been built, the model was an ideal machine in which to study accelerator dynamics experimentally. In an FFAG machine, the acceleration process and the transverse focussing are completely separate functions. This is one of the advantages of the FFAG design. It was easy to study the focussing properties, independently of the accelerating process, and conversely. Initially, Jones and Terwilliger used betatron acceleration to accelerate the electrons. You can see in Figure 1 the rectangular betatron core which links the vacuum donut and focussing magnets. When the magnetic flux through the core is increased, it behaves like a transformer, with the electron beam playing the role of the secondary coil. The induced voltage around the ring accelerates the electrons continuously while the flux is increasing. The rise time is much longer than the time required to accelerate the electrons, so electrons are continually being accelerated from the injector at the inside to the outside radius of the vacuum chamber. Jones and Terwilliger were thus able to achieve a much larger duty factor (ratio of time during which beam is arriving at the final energy to total time) than in a conventional betatron where the magnetic guide field increases in synchronism with the flux in the core, and the beam arrives at the target in a short pulse. The resulting great increase in average beam intensity was one of the advantages of FFAG accelerators. Among the other experimental achievements of the Michigan group were:

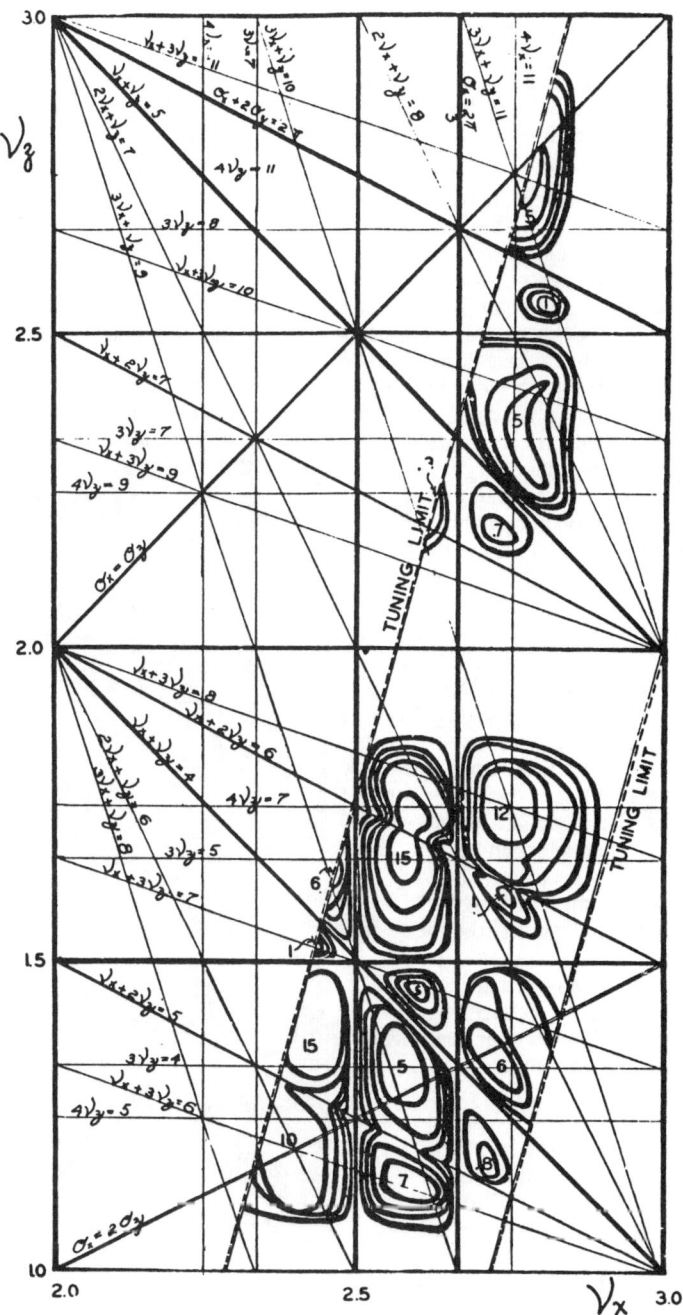

Fig. 2. Resonance Survey - Radial Sector Model

- The invention of a process called rf knockout to measure the transverse oscillation frequencies of the electrons.
- The measurement of beam intensity as a function of transverse oscillation frequencies. Figure 2 shows their results in the form of a contour plot of beam intensity vs. horizontal oscillation frequency v_x (in oscillations per revolution) and vertical frequency v_z. You can see clearly the reduction of beam intensity, sometimes to zero, along the resonance lines where theory had predicted that the beam would become unstable.
- A study of the rf acceleration process, including a demonstration of phase displacement acceleration, a demonstration of beam stacking, and a study of adiabatic capture of the beam into an accelerating "bucket".
- In the Summer of 1956, the Michigan model was disassembled, moved to the MURA lab in Madison, and was quickly back in operation.

The rf acceleration experiments illustrate the advantage of FFAG in studying acceleration dynamics. An rf accelerating voltage was applied across a gap in the vacuum chamber. Betatron acceleration was used to move the beam to an energy intermediate between injection and final energy. The rf voltage was then applied and manipulated in some way. Finally, the beam was again betatron accelerated to a target at the final energy. The signal at the target was recorded on an oscilloscope (Figure 3). The time of arrival is a measure of the electron energy after the rf experiment. Electrons with the highest energy reach the target first. Figure 3 shows a beam stacking experiment. The first trace shows the unmanipulated beam pulse; its arrival time corresponds to the initial energy. The second trace shows the result after one rf accelerating cycle. Most of the beam arrives earlier, showing that it has been accelerated to a higher energy. The remaining traces show the result after four, seven and ten successive acceleration cycles. Beam remaining at the initial energy is picked up on successive cycles and accelerated. However, successive cycles also disturb beam previously accelerated, scattering it in energy and decelerating it by the process of phase displacement to be discussed later.

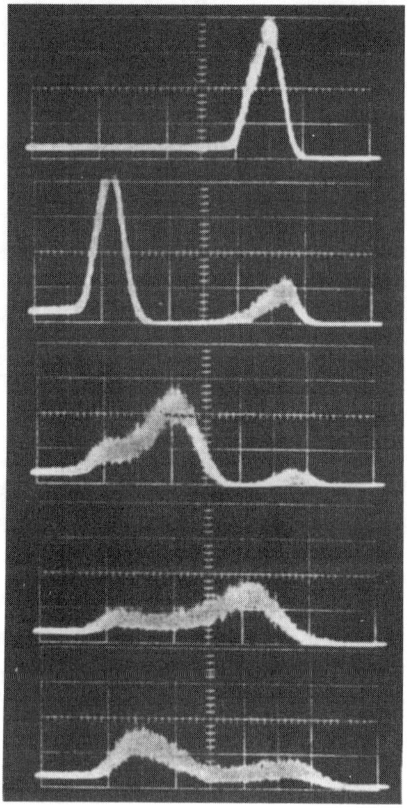

In October, 1954, Donald Kerst invented the spiral sector FFAG focussing principle. In a spiral sector accelerator, the magnet pattern spirals around the circumference as it extends out radially. [See Figure 4.] The focussing comes primarily where the beam crosses the slanted magnet edges. Jackson Laslett and I at first thought that the main advantage of the spiral sector design was that it was so complicated that it was hard to show theoretically that it wouldn't work. Eventually we succeeded in analyzing it, with the result that it would indeed work very well.

Fig. 3. A Beam Stacking Experiment

A spiral sector model was begun in 1956. It was built in Madison by Robert Haxby, Charles Pruett, Ednor Rowe and William Wallenmeyer. The output energy was 150 keV. This was, I believe the first fully computer designed accelerator to be built. Magnet designs were tested by computing the magnetic fields they would produce, and then calculating the electron orbits in the computed fields. Perhaps as a result, when the model was completed in August, 1957, it worked the first time the power switch was turned on. I believe that is also a first. Among the experimental results obtained by the group who built this model are:

Fig. 4. The Spiral Sector Model

- A study of beam intensity versus horizontal and vertical betatron oscillation frequency. [Fig.5].
- The first detailed study of space charge limits in an accelerator. [See below.]
- Discovery and explanation of the window shade instability.
- Extraction of the beam in 1961.

The space charge limit is a limit on the maximum achievable beam intensity set by space charge effects. The group studied these limits both for unneutralized electron beams and for beams neutralized by positive ions created in the residual gas in the vacuum chamber. They measured the maximum allowed beam currents over a range of transverse oscillation frequencies. The results agreed well with theoretical predictions of these effects.

The group observed that when an unneutralized electron beam is injected at an intensity above the space charge limit, some beam is lost and the beam remaining in the accelerator is at the space charge limit. This is to be expected. As the injected beam is increased in intensity, the transverse oscillation frequency is reduced because the mutual repulsion of the electrons is a defocussing effect. When the frequency is reduced until it lies too close to one of the resonance lines in Figure 5, the beam becomes unstable, and electrons begin to be lost. The beam intensity therefore decreases and the oscillation frequency moves away from the resonance.

On the other hand, when an electron beam is fully neutralized, its net charge density is zero, so the mutual repulsion of the electrons is canceled by the positive charge of the (stationary) ions held in the region occupied by the beam. The magnetic field produced by the moving electrons exerts an attractive force, so that the frequency of the transverse oscillations is increased. When the frequency reaches a resonance line, electrons begin to be lost. There are now more than enough ions in the beam to neutralize the electrons. The ions move much more slowly than the electrons, so they remain in the neighborhood of the beam for a while, even after the electrons holding them are lost. As a result, the excess ion attraction on the electrons makes their oscillation frequency even higher, pushing them further into the resonance band. The experimenters observed that as they increased the injected beam intensity under conditions when the beam is neutralized, a limit is reached beyond which the beam suddenly disappears entirely. This effect was dubbed the *window shade* instability, since the beam simply rolls itself up (and out) as more and more of the beam is rapidly sucked into the resonance region by the increasing excess of positive ions.

A disadvantage of FFAG accelerators is the large circumference factor of these machines. Since the

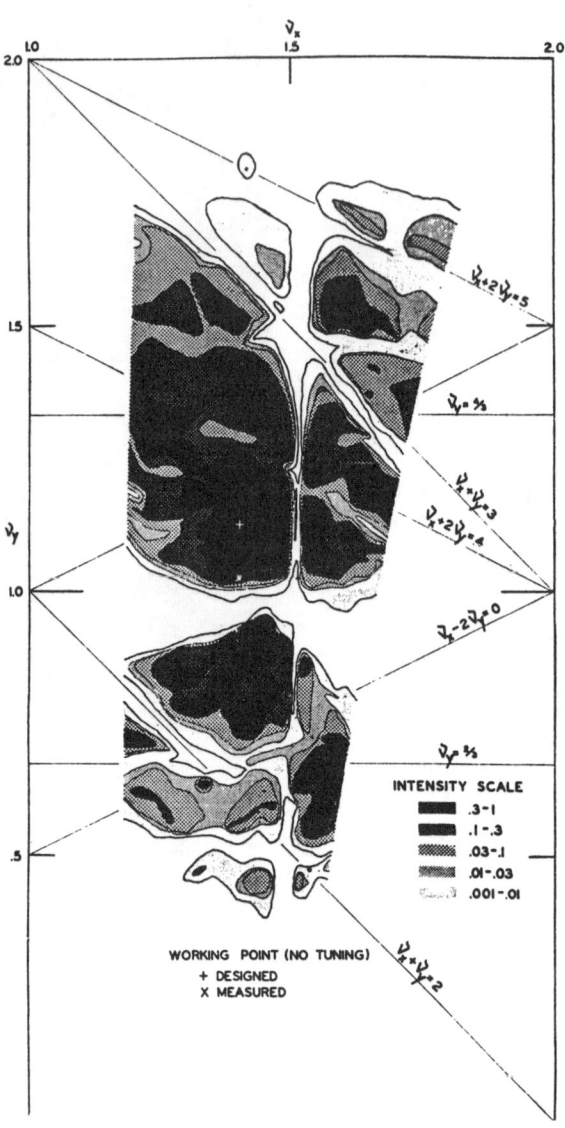

Fig. 5. Resonance Survey - Spiral Sector Model

field in a spiral sector design is always in the same direction, the circumference factor is smaller (about two) than in the radial sector design. This made the spiral sector design very attractive for FFAG cyclotrons. Such a cyclotron can provide intense beams at energies above those which can be reached by conventional cyclotrons. That is because, as also noted earlier by L. H. Thomas, the revolution frequency in an FFAG cyclotron can be kept constant even at energies where relativistic effects would make the frequency in a conventional cyclotron begin to decrease. A number of spiral sector FFAG cyclotrons have been built commercially. This is so far the only practical application of the FFAG principle, except for models used for studying orbit dynamics.

NONLINEAR ORBIT STUDIES

Nonlinear orbit studies were a major activity of the MURA theoretical group throughout the history of MURA. Orbit computations with nonlinear forces were carried out on the Illiac computer at the University of Illinois. The group studied stability limits due to what are now called islands. I remember one night when the computed orbits did not seem to work out correctly, as if the beam did not know exactly when it had arrived back where it started as it went around the accelerator. Somebody eventually discovered that Don Kerst had read the value of π from his slide rule and entered it into the computer. Since the slide rule value was not quite correct, 2π did not carry the particles exactly once around the accelerator.

Later, the orbit computations were transferred to the IBM 704 computer in the MURA laboratory in Madison. The lab was in a garage on University Aveenue. The 704 was the most advanced machine at the time. I believe it must have been roughly comparable in power to an Apple II microcomputer. I base that judgment on my memory of a chess program someone brought in to try on the 704. I remember how long it took the 704 to decide on a move. My Apple II plays chess better and faster, and it did not cost several million dollars.

Jurgen Moser was invited to discuss his theoretical methods of treating nonlinear dynamics. We found a close correlation between what he predicted and what we found in our orbit computations. Several of us became interested in these analytical methods, and used them in studying the various nonlinear resonances that occur in alternating gradient accelerators. We also computed orbits numerically and compared the results with theory. We observed the predicted resonance islands. Similar phenomena are observed in studying the rf acceleration process as I will point out in the next section.

RF ACCELERATION

An idea that arose very early in our consideration of fixed field accelerators was that it should be possible to accelerate successive beam pulses from injection to the final energy, and leave them circulating at the final energy. In this way it might be possible to build up very intense circulating beams at high energy. Donald Kerst mentioned this idea to Eugene Wigner, who immediately got to the heart of the matter by asking "What about Liouville's theorem?" When Don brought this question back to me, I saw that it was the key to the analysis of the process of beam stacking, and indeed to beam handling in general. Liouville's theorem states that the beam particles move in the energy-phase space like an incompressible fluid. This sets a limit to the beam stacking process, since each beam pulse must occupy its separate area in the phase space. One could now calculate the maximum beam that could theoretically be stacked at high energy, given the phase space area occupied by each injected beam pulse. It turned out that if one were careful in capturing and accelerating the beam, so as not to dilute too much the phase space occupied by each beam pulse, one could indeed stack high intensity beams.

Andrew Sessler and I submitted a paper to the 1956 CERN Symposium on Accelerators and High Energy Physics entitled "Methods of radio-frequency acceleration in fixed field accelerators with applications to high current and intersecting beam accelerators.". We analyzed the acceleration and beam stacking processes.

The methods used to study nonlinear transverse oscillations are also applicable to synchrotron oscillations, i.e., oscillations in energy and rf phase. In Figure 6, I show a plot of energy vertically, and horizontally, the phase at which the particle arrives at the rf accelerating gap. This is a phase plot for an FFAG accelerator designed to accelerate particles from a low energy to over 1000 MeV. A point is plotted each time the particle returns to the gap. In this case, the points are seen to lie on closed curves. The single points at 502 MeV and 822 MeV are points where the particle energy is such that the radio frequency is an exact multiple of the revolution frequency, and the phase is such that the voltage is always zero when the particle arrives at the gap. The particle therefore remains fixed in energy and phase, and its synchrotron orbit is a single point. The closed curves around these points correspond to particles oscillating in energy and phase around the central fixed points. Particles on these curves are said to be trapped, and the region occupied by these curves we named "buckets". The name comes from the fact that the points in this space move like an incompressible (two-dimensional) fluid. If the radio frequency is modulated, the fixed points and their associated buckets move up or down in energy, carrying with them any particles trapped in these buckets. Outside the buckets, the curves extend right across the phase plot from zero phase on the left to 2π on the right. A particle on one of these curves is not trapped, but passes through all phases with respect to the rf voltage. Note the small subsidiary buckets between the large ones. They correspond to regions where the motion would become chaotic if the voltage were increased until the two large buckets begin to overlap. An example of chaotic motion is shown in Figure 7, which corresponds to a very high rf voltage. We see the closed curves near the central fixed point, corresponding to stably trapped particles. Farther out are shown two orbits (circles and triangles) in which the points corresponding to successive returns of the particles to the rf gap do not seem to lie on any discernible curve. We called these motions "stochastic". The more popular term today is "chaotic".

One interesting consequence of the fact that the phase space moves as an incompressible fluid is that if you accelerate an empty bucket past a beam initially outside the bucket, the beam suffers an average energy loss proportional to the phase area of the bucket which now lies above it. This process is called *phase displacement*. It is an important factor in limiting the efficiency of beam stacking, in which many injected beam pulses are accelerated successively to the final energy, in order to build up an intense circulating beam. (See Fig. 3 above.) An experiment to demonstrate phase displacement was performed on the radial sector model and is shown in

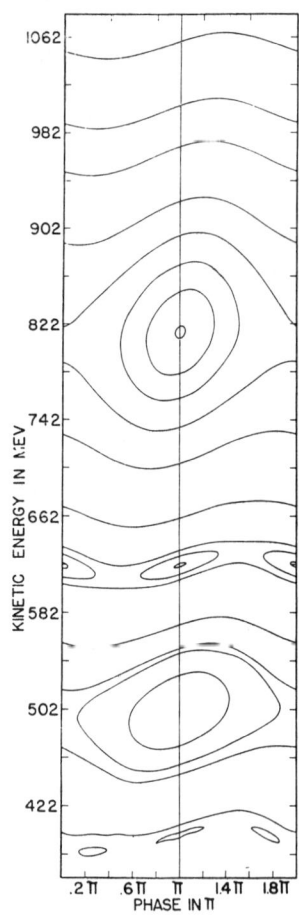

Fig. 6. Motion in Synchrotron Phase Space

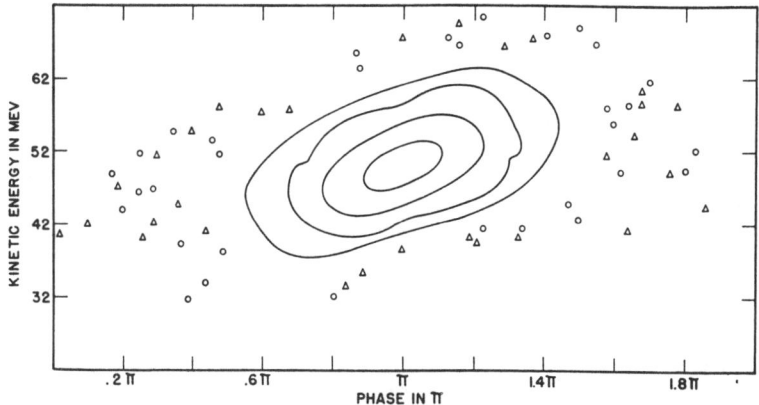

Fig. 7. Chaotic Motion around a Bucket

Fig. 8. The first trace simply shows the second betatron pulse. The second trace shows the beam deposited at the initial energy by the first betatron pulse, (as it strikes the target after the second betatron pulse.) In this experiment, the rf was turned on at a frequency corresponding to an energy above the initial beam energy, and the frequency was then increased to correspond to an even higher energy. As a result, the initial beam is not captured. We then do the experiment again, this time turning on the rf at a frequency synchronous with the initial energy of the beam. We see (third trace) that at the end of the accelerating cycle, most of the beam is at a higher energy, with a little beam left behind that was not trapped by the rf bucket. The fourth trace is an experiment where the rf is turned on at a frequency below that of the beam, and then the frequency is increased to a level above that of the beam. We see that no beam was captured, but that the energy of the beam decreased a little as the accelerating rf bucket passed by. This is the phase displacement process, predicted theoretically and observed for the first time.

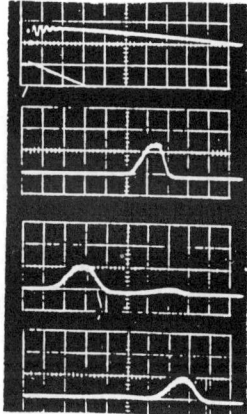

Fig. 8. Phase Displacement

COLLIDING BEAMS

If one can stack high intensity beams, then it becomes practical to think of colliding beam experiments. Many people had realized that one can reach a much higher effective energy by letting two high energy beams collide than one gets by letting a high energy beam from an accelerator strike a stationary target. When a particle of high energy E strikes a stationary particle, then according to the equations of the theory of relativity, the center of mass energy, which is what counts in a reaction between the particles, is proportional to the square root of E. Thus a four-fold increase in accelerator energy results only in doubling the useful center of mass energy. On the other hand, if two particles each of energy E collide head-on, the center of mass energy is $2E$. In this case, all the energy is available as useful energy.

Unfortunately, the particle density in even the most intense accelerator beams available heretofore is literally trillions of times smaller than in a solid target. The reaction rates in colliding beams would therefore be trillions of times smaller than when a beam hits a stationary solid target. Until 1956, the reaction rate one could expect if one tried to make two accelerator beams collide was much too small for any practical experiment. With beam stacking, however, calculations showed that one could achieve beam intensities high enough to make colliding beam experiments practical, although the reaction rates were still much smaller than with beams striking fixed targets. The gain in center of mass energy made colliding beams now an attractive possibility. The highest energy reactions are now routinely studied using colliding beams. The late SSC (Superconducting Super Collider) is the most ambitious high energy facility so far proposed.

I remember being asked to give a seminar at the University of Illinois shortly after we had shown that colliding beams might be practical. As soon as I mentioned colliding beams, the entire audience burst out laughing. I was somewhat taken aback. Colliding beams might have been a far out idea at that time, but it was a serious proposal. I was told later that the week before, at a Physics Department picnic, Jerry Kruger and Don Kerst had faced each other firing pea shooters and tried to make the peas collide.

NEGATIVE MASS INSTABILITY

The revolution frequency in an alternating gradient accelerator increases with energy at low energies. This is because the velocity increases with energy. At high energies, as the velocity approaches the velocity of light, it increases only very slowly with energy. However, for a given magnetic field strength, the orbit radius increases with energy. As a result, at very high energies, the revolution frequency decreases with energy. There is a transition energy which marks the dividing line between low energies where the frequency increases with energy, and high energies, where it decreases with energy.

Carl Nielsen, while at CERN, I believe in 1958, realized that above the transition energy the inter particle forces behave oppositely to what one would normally expect. Charge particles of like sign repel one another. Above the transition energy a particle which gains energy slows down, and conversely. The result is that the repulsion between two beam particles causes them to move closer together! The one in front (as they move around the accelerator ring) gains energy from the repulsion, and slows down; the one behind loses energy, and speeds up. This can cause an instability, in which particles tend to collect in clumps around the accelerator. This is called the *negative mass* instability, because particles behave as if they had a negative mass -- they tend to accelerate in the opposite direction to the applied force.

In the mid-nineteenth century, James Clerk Maxwell won the Adams prize for an essay explaining the rings of Saturn. Maxwell pointed out that satellites circling a planet move more slowly as they gain energy, because they move to an orbit of greater radius, and conversely. Thus these particles behave like particles above the transition energy in an alternating gradient accelerator. This analogy was pointed out to me by Carl Nielsen. Without this insight, one might have dismissed the idea that Saturn's rings consist of dust or rocks, because one might think the gravitational force between them would cause them to collect in larger clumps. But because of their effective negative mass, the attraction between the rocks keeps them apart! Conversely, in the negative mass instability for charged particles, the repulsion between the particles causes them to clump together.

After Carl wrote me about the effect, he and Andrew Sessler and I worked to develop the theory. Our work resulted in a paper presented at the 1959 CERN Conference. The effect was observed in the Cosmotron. The Cosmotron is a "weak-focussing" accelerator, i.e. it does not use the alternating gradient principle. In a weak focussing accelerator, the transition energy is zero, and the beam is always above transition. The negative mass instability has been seen in many machines, and is now one of the effects one must take into account in designing accelerators. It has also been seen in toroidal plasma devices.

MURA IN RETROSPECT

Soon after the invention of FFAG, it became the main focus of the activity of the MURA group. It was a new idea, and its advantages seemed at the time more important, and its disadvantages less important than they later turned out to be, relative to other accelerator designs. During the period from 1955 to 1962, MURA submitted a number of proposals to the Atomic Energy Commission for FFAG accelerators in the 10 to 20 GeV range, including a 15 GeV two-way colliding beam accelerator, to reach a center-of-mass energy equivalent to that for a single 540 GeV beam colliding with a stationary target. Curiously, the President's Science Advisory Committee panel studying this proposal concluded that very high energies were probably not very interesting. In hindsight, that seems a strange conclusion, but at the time, it was not believed that there could be anything but very broad resonances in particle reactions at very high energies, because particles in high energy states should have very short lifetimes. The panel concluded that very intense beams at 15 GeV would be useful to study reactions with very small cross sections at center of mass energies of a few GeV. It recommended that first priority be given to the Stanford two-mile linac, and that next priority be given to a high intensity proton accelerator. As a result, MURA abandoned colliding beams, and on March 30, 1962 submitted a proposal for a 10 GeV high intensity FFAG accelerator. It was to have an intensity of 2×10^{14} to 2×10^{15} protons per second, or about 2000 times the intensity from the alternating gradient synchrotron at Brookhaven. Although the AEC paid the costs of preparing them, none of these proposals was ever funded. In retrospect, that was probably a wise decision. The machines were carefully designed, and the designs were based on working models built by MURA. They would have worked as proposed, but they would have been monster machines, using vast amounts of iron, copper, and money. We later learned how to build much less bulky and less expensive machines to accomplish the same purposes.

The FFAG principle studied by MURA has had little direct application, so in a sense it was a mistake for MURA to concentrate on FFAG accelerators. However, the MURA studies led to many important developments in accelerator technology, and had a significant impact on present day accelerator designs. Among the MURA accomplishments were:
- Colliding beams - The MURA group demonstrated that they could be practically achieved.
- Storage rings for colliding beams - In the Spring of 1956, as a result of the MURA studies, Lichtenberg, Newton, and Ross at Indiana University, and independently O'Neill at Princeton proposed to stack intense circulating beams in fixed energy rings. Such rings would have fixed fields and alternating gradient focussing, but they would be built like conventional synchrotrons, with small apertures, and therefore would be much less expensive than FFAG machines. A conventional accelerator would accelerate particles to full energy before injection into the storage ring.

- Contributions to the theoretical and experimental study of nonlinear dynamics in accelerator beams.
- Careful design of the entire accelerator, particularly of the magnets, using computers to calculate magnetic fields, and orbit properties. This approach was particularly emphasized by Donald Kerst, who led the MURA technical group during much of its history. Earlier, Kerst and Serber had published a careful theoretical analysis of transverse oscillations in a betatron. Such oscillations in any accelerator are now called betatron oscillations, in recognition of the work of Kerst and Serber.
- Ideas, analysis, and experimental studies of the manipulation of beams by manipulating the frequency and voltage of applied rf fields. These processes are much easier to analyze and to study experimentally in FFAG machines.
- Beam stacking.
- Discovery of the negative mass and other instabilities.
- Experimental verification of theoretical estimates of space charge limits on beam intensities.
- Invention of the spiral sector cyclotron.

Over 600 MURA reports were written describing the work of the MURA group, plus many published papers.

1959 MURA SUMMER STUDY

In a sense, MURA was responsible for its own demise. In the summer of 1959, MURA hosted a study on the Design and Utilization of High Energy Accelerators. Figure 9 is the cover page of the report on the study. MURA people will recognize the standard MURA report cover. Figure 10 is a list of the participants. You will recognize many well-known names among them; sadly, a number of them are no longer with us. The participants divided into groups to study the design and the experimental utilization of high intensity and high energy accelerators. The study produced a number of interesting reports [Fig.11].

Perhaps the most important report from the study, because it sounded the death knell of high energy FFAG accelerators, was a report by Matthew Sands on cascade synchrotrons [Fig.12]. Sands studied a cascade synchrotron accelerator whose parameters are shown in Figure 13. Beam is accelerated in a synchrotron of intermediate energy (10 GeV). It is then extracted and injected into a high energy synchrotron, where it is accelerated to 300 GeV. The design of the Fermilab accelerator was remarkably similar to Sands' original proposal.

Up until then, it had not been clear that beams could be efficiently transferred from one ring to another. If not, then neither cascade synchrotrons for reaching high energy, nor storage rings for colliding beams would be practical. The only way to get high intensities and high center-of-mass energies would have been FFAG.

Sands in his report showed that a transfer of beams was probably practical, and that the machine he proposed would perform satisfactorily. Moreover, the cost was only $75,000,000, comparable to the cost of a 10 GeV FFAG machine. The center-of-mass energy for 300 GeV protons colliding with stationary protons would be nearly 25 GeV, the same as would be achieved with colliding beams at 12.5 GeV. If it worked, FFAG was dead! It did, and all high energy synchrotrons built since have used the cascade principle. Colliding beams, however, were not dead. By transferring beams from high energy cascade synchrotrons to high energy storage rings, and using beam stacking in the storage rings to build up intense circulating currents, one can now contemplate building machines to reach several thousand GeV center-of-mass energy.

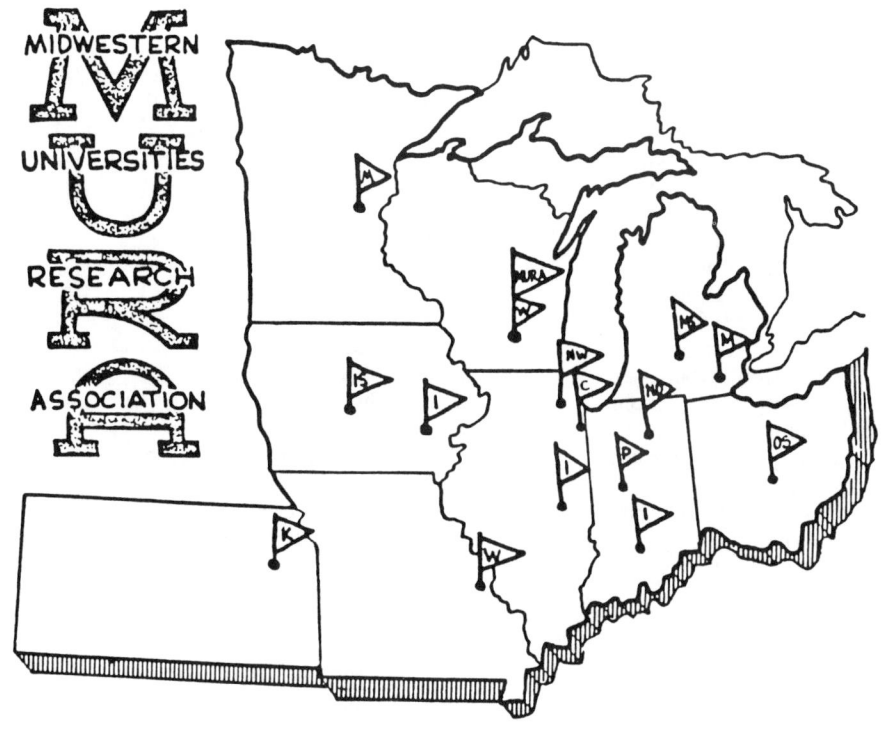

1959 MURA SUMMER STUDY REPORTS ON

DESIGN AND UTILIZATION OF HIGH-ENERGY ACCELERATORS

REPORT **NUMBER 506**

Fig. 9. Cover Page from the 1959 MURA Summer Study Report

LIST OF PARTICIPANTS

R. K. Adair, Brookhaven
E. S. Akeley, Purdue
J. P. Blewett, Brookhaven
M. H. Blewett, Brookhaven
H. G. Blosser, Mich. State
H. Bruck, Saclay
N Christofilos, Livermore
R. L. Cool, Brookhaven
E. D. Courant, Brookhaven
H. R. Crane, Michigan
D. H. Dalitz, Chicago
  W. Fulbright, Rochester
E. L. Goldwasser, Illinois
M. L. Good, Berkeley
C. L. Hammer, Iowa State
A. O. Hanson, Illinois
R. H. Hildebrand, Argonne
N. Horwitz, Berkeley
L. W. Jones, Michigan
D. L. Judd, Berkeley
  Kitagaki, Princeton-Penn
P. G. Kruger, Illinois
L. J. Laslett, Iowa State
P. Lax, New York
D. B. Lichtenberg, Mich. State
D. Meyer, Michigan

C. G. Allan Mitchell, Indiana
R. F. Mozley, Stanford
C. Mullin, Notre Dame
C. E. Nielsen, Ohio State
G. K. O'Neill, Princeton
T. R. Palfrey, Purdue
R. F. Post, Livermore
B. Richter, Stanford
A. Roberts, Rochester
K. W. Robinson, CEA
M. Sands, Cal. Tech.
M. Schein, Chicago
A. Schoch, CERN
A. M. Sessler, Ohio State
F. C. Shoemaker, Princeton
G. Tautfest, Purdue
L. C. Teng, Argonne
K. M. Terwilliger, Michigan
R. Thom, Sao Paolo
A. V. Tollestrup, Cal. Tech.
W. D. Walker, Wisconsin
A. Wattenberg, Illinois
T. A. Welton, Oak Ridge
T. Ypsilantis, Berkeley
D. J. Zaffarano, Iowa State

Fig. 10. 1959 Summer Study Participants

CONTENTS

INTRODUCTION

    MURA-506 Remarks on Elementary Particles and Their Interactions.
D. B. Lichtenberg

UTILIZATION OF HIGH-INTENSITY MULTI-BEV BEAMS

    MURA-469 Summary of Discussion of the Physical Interest in Higher Intensity.
R. L. Cool, A. Wattenberg and T. Ypsilantis

    MURA-470 Desirable Beam Characteristics of High-Intensity Accelerators.
R. K. Adair, D. Meyer, B. Richter and G. Tautfest

    MURA-473 Notes on the Session on the Experimental Uses of High-Intensity Accelerators. A. Roberts.

    MURA-471 Energy Considerations of High Intensity Accelerators. W. D. Walker

    MURA-472 Beam Separators. B. Richter, M. L. Good, D. Meyer
    MURA-576 Proposed Radiofrequency Separator. B. Richter

ULTRA HIGH ENERGIES

    MURA-479 Utilization of Very High Energies. E. N. Goldwasser, C. L. Hammer, L. W. Jones, D. B. Lichtenberg, C. Mullin, M. Sands, A. V. Tollestrup and W. D. Walker

    MURA-499 Background in the Neighborhood of a Colliding Beam Region.
D. B. Lichtenberg and L. W. Jones

HIGH-INTENSITY AND ULTRAHIGH ENERGY ACCELERATORS. Design and Cost Studies.

    MURA-467 A High Repetition Rate 15 Bev Alternating Gradient Synchrotron.
M. H. Blewett, E. D. Courant, H. W. Fulbright, F. C. Shoemaker, A. V. Tollestrup, T. A. Welton

    MURA-478 The Cost of a 10 Bev Proton Linear Accelerator. R. F. Mozley

    MURA-468 Cost of a 10 Bev Scanning Field AG Synchrotron. T. Kitagaki

    MURA-460 Cost Estimate Comparison of Three Multi-Bev FFAG Accelerators.
K. M. Terwilliger

    MURA-465 Ultra High Energy Synchrotrons. M. Sands

PRINCIPLES AND NEW IDEAS IN ACCELERATOR DESIGN

    MURA-462 High Current Effects in FFAG Accelerators. F. T. Cole, R. O. Haxby, L. W. Jones, F. E. Mills, C. E. Nielsen, A. M. Sessler, K. R. Symon, K. M. Terwilliger

    MURA-475 Multiple Frequency Accelerators. K. W. Robinson

PARAMETERS OF 50 MEV PROTON LINEAR ACCELERATOR AT BROOKHAVEN

LIST OF PARTICIPANTS

Fig. 11. List of MURA Summer Study Reports

MURA 465
TID-4500 (15th Edition)
UC-28

MIDWESTERN UNIVERSITIES RESEARCH ASSOCIATION*

2203 University Avenue, Madison, Wisconsin

ULTRA HIGH ENERGY SYNCHROTRONS

Matthew Sands
California Institute of Technology

June 10, 1959

ABSTRACT

A discussion is given of the technical feasibility and economic scale of pulsed AG synchrotrons with maximum energies of 100 and 300 BeV. The 100 BeV machine is similar to a scaled Brookhaven (30 BeV)AG synchrotron, while the 300 BeV machine consists of two synchrotrons in cascade, the beam being transferr3ed to a small aperture second stage at 10 BeV. It is concluded that the cost of either the 100 or 300 BeV is of the same order as a 10 BeV high-intensity FFAG accelerator.

---

*
AEC Research and Development Report. Research supported by the Atomic Energy Commission, Contract No. AEC AT(11-1)-384.

Fig. 12. MURA Report 465 - Abstract

MURA-465

TABLE II
PARAMETERS OF THE 300 BEV ACCELERATOR

	Stage I	Stage II
Injection Energy	50 MeV	10 BeV
Final Energy	10 BeV	300 BeV
Injection Field	300 gauss	300 gauss
Radius	30 meters	1000 meters
Aperture		
Radial	14 cm	5 cm
Vertical	6 cm	1.5 cm
Focussing		
ν	~3	~100
β	60 meters	60 meters
Acceleration time	1 sec	1 sec
Volts/Turn	7 keV	6 MeV
$\delta f/f$	2/3	1/200

Fig. 13. Parameters for the Cascade Synchrotron

THE END

The MURA laboratory was closed in 1967, and the facility was sold to the University of Wisconsin. The University established the Physical Sciences Laboratory at the site. Fred Mills, then Director of the MURA laboratory, became the first Director of PSL. The lab provides services to University departments when they have projects too big to undertake by themselves. The lab also continued the MURA experimental program in accelerator physics, under Ed Rowe.

The MURA corporation was dissolved in the 1970's.

ANDREW SESSLER'S LBL DIRECTORSHIP

To understand Andrew Sessler's role as the Director of the Lawrence Berkeley Laboratory from late 1973 to 1980, it is necessary to recall the troubled situation the Laboratory was in during the early 1970s.

In the history of science few scientific laboratories have achieved the dominant position that Ernest Lawrence's Radiation Laboratory did in the nineteen thirties up through the two decades after World War II. Lawrence, the pioneer, led the way with his cyclotrons in the prewar period and with his 184 inch cyclotron and Bevatron after the war. These machines attracted a group of young geniuses whose revolutionary discoveries in physics, chemistry, medicine and other sciences brought to them more than a half dozen Nobel Prizes. It is often said that "Big Science" was created by Lawrence and his Laboratory . The Radiation Laboratory was Mecca for nuclear and particle physicists. In Luis Alvarez's phrase it was " A Laboratory like no other". Lawrence's premature death in 1958 was a serious blow , but his protégé, Edwin McMillan, became Director and maintained his traditions and continued the outstanding research program for many years.

But all great scientific institutions have their life histories and there was no way that Berkeley's dominant position could be maintained indefinitely. Indeed, Lawrence's success and his optimistic, exuberant spirit inspired the drive for design and construction of bigger accelerators at many places around the world. The time was approaching when the Radiation Laboratory , now renamed the Lawrence Berkeley Laboratory, had to adjust to a new role in its traditional areas of expertise in a world where many laboratories had accelerators delivering nuclear particles of higher energy. By the early 1970s the budget and staff size were declining and morale was slipping.

There were other changes afoot in the early 1970s which required some urgent attention. The pattern of funding from Washington was

© 1996 American Institute of Physics

changing toward research and development in new energy technologies to help the nation cope with problems caused by its sharply increasing need for energy, particularly in the form of liquid fuels, by decreased domestic production of petroleum, by increased dependence on Middle East oil, and by huge increases in the price of OPEC oil. Washington priorities were shifting toward alternate energy technologies, toward energy conservation and to the reduction of environmental and health affects of energy technologies. These concerns were soon to lead to the abolition of the US Atomic Energy Commission, the formation of the a new agency called the Energy Research and Development Administration (ERDA) and a few years later, during the Carter Presidency, by the abolition of *that* agency and its replacement by the Department of Energy (DOE).

It is true that Andy Sessler was not the only one who had his worries about his laboratory's preparation for the new world, but he was one of the few who did something about it. In partnership with Dr. Jack Hollander and with the blessing of Director McMillan he started discussions with members of the Berkeley Faculty about multidisciplinary programs they might mount to help solve the non-nuclear energy problems of the nation in partnership with LBL scientists, making use of the superb engineering and back up resources "on the hill". It was this activity that brought Sessler's name to the attention of the university state-wide administration and the Board of Regents when a new Director was needed after Professor McMillan announced in 1972 his intention to retire. Of course, Sessler's high intellect and his record of achievement as a scientist were also an essential requirements for consideration because no one without them could dare to assume overall leadership of the pack of tigers, including several Nobel Prize winners, who led the research groups of LBL.

Upon assuming his new position late in 1973 he did greatly broaden the scope of LBL's programs by making use of the impressive pool of talent in the faculty and graduate students of one of the greatest institutions of higher learning in the world situated immediatedly adjacent to his national laboratory. None of the other national laboratories had the possibility of such synergistic close coupling to academia as did LBL and Sessler exploited this to the full. New programs were added in solar power, geothermal power, energy conservation, building energy science, electrochemistry, material science, environmental science, geology, geomechanics, reservoir engineering, and many others. Similar changes were going on in the other national laboratories, but the LBL/UCB collaboration had its very special characteristics. At any one time 600 or more graduate students were doing PhD thesis research in 14 academic departments. It is unfortunate that this short review precludes the possibility of describing the numerous highly-interesting developments in non-nuclear as well as in fusion energy that came out of these programs.

Although these new programs represented a huge expansion of the research interests of the Laboratory, the traditional nuclear fields were not neglected. Many LBL physicists traveled to other laboratories to do research, but these same physicists continued working at LBL with the superb engineering and detector development groups to design and build particle detector systems, most of them with large and complex configurations. These were later transported to other accelerator sites where they played a crucial role in major discoveries. Particle physics has remained alive and well at LBL but its focus has been changed to fit the changed world. Sessler encouraged the scientists and engineers involved with accelerators to follow their dream of linking the Superhilac, a lower-energy heavy ion linear accelerator, with the Bevatron to produce heavy nuclear particles of relativistic energy. This Bevalac development inaugurated the whole new science of relativistic heavy ion nuclear physics. These Berkeley achievements brought new fame and and spawned related programs in Germany and at CERN and an exciting prospective program at Brookhaven. The Bevalac also supported a unique program in medical science for the treatment of disease using the unique properties of relativistic heavy ions.

The above can only hint at the many changes in scientific research under Andy Sessler's leadership; it is necessary to save some space to discuss his influence on organization and laboratory sociology.

The Laboratory he inherited from Lawrence and McMillan had an informal and very loose style with many group leaders of the warlord persuasion. This was one of the charms of the place and it worked brilliantly as long as the staff was not too large and a strong charismatic leader was around to deal with the big problems. But in the new world this would not do. Andy also had an informal personal manner but believed in the necessity for clear organizational relationships. He clarified the divisional organization. He insisted that the Division heads meet on a regular basis to serve as his council of advisors on labwide policy issues. Each had the title of Associate Director of the Laboratory and he wanted each one to take that title seriously and accept responsibility for the overall health of the institution. He actually published an organization chart! He instituted the policy of limited appointments for Associate Directors. He established rules regarding contacts with government officials. He organized formal Outside Users Groups to help experimentalists from afar to have a voice in the operating policies of the LBL accelerators and to give them a fairer shake in the allocation of beam time. His Administration prepared a personnel manual with job classifications, salary schedules, job benefits information, travel rules, budget rules, etc. This was a vast change for "Rad Lab" veterans and it was not received with uniform exultation and praise. No scientist likes this kind of bureaucracy and boilerplate but when the staff grows to thousands there is chaos without it.

The faculty/ staff scientist duality caused some prestige and social problems which were dealt with by creating Senior Staff titles for top level scientists and engineers, with formal procedures similar to campus rules for nomination and award of the title. The multidisciplinary culture of LBL , which was greatly different from the UCB campus research practise, also tended to blur the hierarchical relationships.

Andy introduced the idea of quality review throughout the Laboratory. He appointed panels of distinguished outside professionals to review the programs of each division on an annual basis with confidential reports to the Division Head and to the LBL Director. These reports resulted in pruning of weaker programs, sharpening of goals, introduction of new goals, etc. This rigorous process strengthened the Laboratory.

Throughout all his six plus years Andy maintained his informal personal style. He kept the respect of his old and several new tigers for his intelligence, his honesty, his lack of search for personal gain, and his commitment to maintaining this institution as one of the world's premier institutions for fundamental research and development.

In summary, Andrew Sessler took over a laboratory with a declining budget and workforce, with an uncertain future in the fields it had dominated during its glory days, and without a clear plan for involvement in the scientific research in the technologies required for the impending energy crisis. He led the laboratory to a revised but still quite significant role in its traditional fields while greatly broadening its programs to recast it as a multipurpose laboratory with national program leadership in many important areas. We can all be grateful for his inspired leadership at a time when that leadership was sorely needed.

This week we are participating in a Symposium on Beam Physics to honor Andrew Sessler for his contributions. Many of these contributions came in the years after his term as LBL Director. This is quite remarkable. I have known many people who swore they were happy to lay down the burdens of management so they could return to a career of total commitment to their profession. Very few actually matched their professed resolve with action. We are pleased to express our admiration for Andy in his impressive successful reincarnation as a working, productive scientist.

Some remarks prepared by Earl K Hyde who served as Director Sessler's Deputy. These remarks were prepared for delivery at the banquet, December 5, 1993 during the Sessler Symposium on Beam Physics.

Andrew Marienhoff Sessler, *Physicist*
A Student's Guide to the Couplings of a Beam Physicist with His Environment

(Uninvited Paper)

AMS Collaboration*
Lawrence Berkeley Laboratory, University of California, Berkeley CA 94720

Andy's life is reviewed briefly, with emphasis on what is known. Prospects for the future are discussed.

I. Caveat
"Freunde! Nicht diese Tone!..."

In this volume, *others* tell of a physicist and the field he fathered, **beam physics**, or perhaps, co-fathered. They write not to glorify the field, since it's just physics, they write to chronicle an era---as a little-known poet has written,

To recount the days
when Man built beacons beneath the Earth,
to raise bare the forbears of Matter,
and called these machines
Cyclotron! Synchrotron! and Linac!
accelerators *all*
behemoths of Pipe and Pump
that lay bare too
the genius of their calculating crew...

or words to that effect.
 I can tell you no such epic tale of clashing pipe and whooshing flange, only a brief annotation, transcribed in haste from black & white and uncertain memory---I had not even met this man, **Andy**, when the Doctor came to him in 1980 and told him the end was near---but let's start at the beginning.

---

*Spokesman: D.H. Whittum.

II. It's A Wonderful Life
"I sing of Andy glad and big..."

Andrew Marienhoff Sessler was born in Brooklyn, NY on 11 December 1928, and his parents, both teachers, moved to **Queens** when he was two, where Andy thought it best to join them. Andy was followed in 1931 by a brother **Roger**, and by a sister **Joyce** four years after that. His parents were delighted to see that Andy would not be an only child.

A. Family Values

Andy has always been reticent about his involvement with this family, but we do know that his brother Roger at least was a success, as was his sister Joyce. At first thought to be a ne'er do well, not interested in intellectual things at all, Roger eventually made a fortune in the trailer park business and retired at age 40. Joyce was also quite successful; she married a businessman who lived in Maine. They are now retired in Florida, having raised three children.

Andy began his formal training, in kindergarten at **PS#117** in Jamaica, Queens, blissfully unaware of his destiny. He climbed out of the PS system in his teens going to Forest Hills High School, also in Queens, where he won the **Westinghouse** talent search at age 15, the first of many from Forest Hills. His parents were ecstatic, thinking they would get a free home appliance. During this time Andy and his best friend **Dave Kera** took many long and arduous bicycle trips together, before they got their drivers' licenses.

B. A Setback for Biology

Andy applied to **Harvard**, a college in Cambridge, Massachusetts, but after the Westinghouse many other colleges wrote to him, asking him to attend. Passing up some good scholarships and ignoring his mother's advice he went to Harvard, reknowned for its excellent libraries. Once in Cambridge he joined the Kirkland House freshman football team, but after a careless encounter with a goal-post, moved on to individual athletics: swimming, rowing, and bike-riding. He also decided to pursue academic studies, first Biology, and then, after the notorious "frog incident", transitioned to **Math**, which he had heard did not involve live dissections.[1]

It was at Harvard that Andy had his first three or four dozen of his many brushes with greatness. For example he met Bryce Seligman (later Bryce DeWitt), who took time in the evenings, on his own initiative, to lecture undergraduates on general relativity among other topics. Actually Bryce had tried doing this for students in grades K-12, but parents complained about his now-famous "belt-demonstration".

Andy also encountered an obscure mathematician who evidently had to work from notes to give his lectures. "What a crumby mathmetician!" the students often cried. "Reading from notes! Ha!", and as the more mature Harvard undergraduates are inclined to say "Hsssss!". But as the days rolled by the notes became fewer and fewer, until finally there were no notes at all. Eventually these young geniuses realized the mathematician had merely been learning English, over the first couple of months; he was Assistant Professor Henri Cartan, who, unusually for Harvard, won tenure long before he retired.

In his summers at Harvard Andy worked at a number of summer camps as Nature Counselor and Waterfront Counselor. During this time, at Phillips Brooks House he met Reverend **Mario Cestaro** who ran a Baptist "settlement house" in Boston's North End. Andy later worked for Mario as Waterfront Supervisor at a two-week church summer camp and, all in all, no one was seriously injured, in fact Andy saved several kids who evidently had never seen livestock before. Andy and Mario still correspond.

C. Simply for the Money

Graduating from Harvard in 1949 Andy went on to **Columbia**, in New York, where he spent some time "catching up" on basic physics. It was in a course in Heat and Thermodynamics taught by Isidor Rabi, that he met a girl named **Gladys**, whom he married in 1952. After writing his thesis, and after a period of nail-biting, as it was carefully read by Norman Kroll, Gladys was relieved to see Andy earn his Ph.D. in 1953.

After Columbia, Andy went on to **Cornell**, on an NSF postdoc to study with Hans Bethe. At Cornell he was offered an assistant professor position at **The Ohio State University**---and Bethe

---

[1] Regarding the notorious "frog incident," see Boston Globe editions from December 1945.

recommended he take it and not do a second year of post-doc. "Simply for the money" Andy allows and Bethe is silent.[1]

D. The Untenured Years

It was during this period that Andy began his work on **atomic and nuclear structure physics**,[2] a hot topic at that time, cluminating some years later in a paper coauthored with Gordon Baym[3]---a paper still being cited today. Andy also pursued over the next decade an interest in **liquid helium**,[4] predicting the **superfluidity** of helium-3,[5] fifteen years before it was observed. In the years to follow Andy would go on to write over twenty papers in this area, the works on the hyperfine structure of He-3 being perhaps the most often cited.

During this time of vigorous professional activity, Andy also had an active home life. His first son **Dan** was born in 1954, his second son **Jon** was born in 1956 and his daughter **Ruth** was born in 1957. A typical day in the Sessler home began with Andy making breakfast for the family, often getting up earlier than the rest to bake scones and biscuits. Evenings were the time for a very leisurely dinner with much good conversation, and a chance to recover from breakfast. Andy also thought of many ingenious games to play, like *"Colliding Beams!"* and *"Monitor the Sun for Neutrinos!"* Gladys would step in when things got too rough.

During the period 1955-56 Andy was "on leave with **MURA**",[6] as his c.v. notes cryptically. "What was MURA?", one wonders. If you should ever dare to ask this of one of the old-timers, be sure to bring a bag lunch. MURA was a great spirit and quite young people, they will tell you, all led by Don Kerst, many years their senior and light years ahead in maturity and judgement. Essentially no one at MURA was an "accelerator physicist," since that field was just being invented. But all had some knowledge of equations.

---

[1] Assistant Professor salaries at Ohio at that time started at $5,400 per year. (Source: Funk & Wagnalls Pocket Guide to Assistant Professor Salaries Throughout History).
[2] Pubs. #1,2,3,4,5,9,10,11,26,27,61 in the c.v., elsewhere in this volume.
[3] Pub. #26.
[4] Pubs. #13,14,15,16 18,19,23,37.
[5] Pub. #13
[6] Midwestern University Research Association.

At Illinois, from the autumn of 1955 to the next autumn, the young physicists worked in a house converted into offices. The room for Jackson Laslett and Andy was so small, we're told, that they had to take turns sitting down. They quickly became close friends. Down the hall were Frank Cole and Tihiro Ohkawa. When the group moved to Madison in June, 1956 Jackson and Andy were left behind to care for the ILLIAC, they were lead to believe. They had one secretary so Jackson and Andy felt it their duty to keep her busy, and wrote about one report a day, and several *errata*.

On Mondays they had to go to Madison to fill the Group in on the week's results and to get new problems. For this purpose they used the University of Illinois airplanes which had four seats and one prop. The two young physicists would change seats so that each got a turn to sit up front with the pilot and offer directions. Going to Madison was fine, but in the Mid-West afternoons thunderstorms developed just about every day, and coming back they would dodge storms and get quite tossed around, possibly the beginning of Andy's fascination with **turbulence**.[1] Back in Urbana the field would be dark, the pilot would buzz the landing strip, and someone would wake up and turn on the lights so that the intrepid crew could make some semblance of a landing. There were occasional close-calls of course, since sometimes other folks in the neighborhood would wake up and turn on their lights too.

The ILLIAC was the biggest computing machine in the world at that time, *i.e.*, the CPU equivalent of a modern Casio wristwatch. It had 1024 memory slots and was programmed in fixed point. Paper tape and photo cell was the input, punched paper tape the output. Storage was in big dark TV-like screens near the ceiling. Andy *et al.* "got the ILLIAC" one evening a week and would all gather there and operate in real time. Using the symplectic maps they had just invented, they typically ran 40,000 turns---and then ran them backwards to be sure they had no truncation problems. They never saw instability, of linearly stable motion, but the length of the run was too small to merit publication. All of this was before the Kolmogorov-Arnold-Moser Theorem; in fact Moser was Ernie Courant's brother-in-law, so they invited him out and this may have been what started Moser's interest in non-linear phenomena.

---

[1] Pub. #69.

Thus, still not thirty years of age, Andy began his more famous works in **rf theory and colliding beams**.[1] Contrary to the impression you might get in reading the now almost biblical 1956 CERN Symposium,[2] Andy did not attend. In fact the paper by Keith Symon and Andy was not even accepted, but Don Kerst insisted and gave up his slot to them. This paper with Keith, the introduction of the "bucket"-concept, is the single most frequently cited paper in the accelerator literature, with references appearing throughout the years in many other venues as well, in works on the applicability of the third integral of motion, chaos, stochasticity and reconnection in Hamiltonian systems, merging tokomaks, variational principles in classical mechanics, and diffusion in the presence of nonlinear resonance. Keith Symon describes this period in greater detail elsewhere in this volume.

At this point in the chronicle, as Andy begins to develop a professional life, you may find it helpful to have a graphical overview as seen in Fig. 1, depicting in histogram fashion publications by year and type, with annotations for certain pivotal events. For further detail on the various categories, the interested reader can consult the c.v. in this volume, or try anonymous ftp to sessler@lbl.

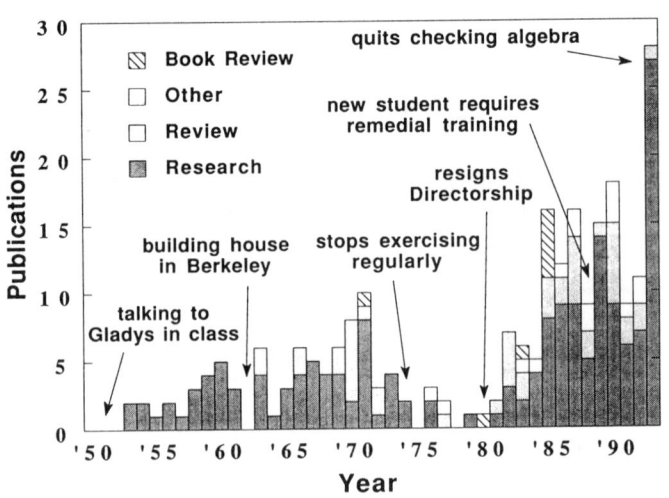

FIG.1 A histogram of Andy's publications by year, not including works in watercolor.

---

[1] Pubs. #6,7.
[2] *Proceedings of the CERN Symposium on High Energy Accelerators and Pion Physics* (CERN Service d'Information, Geneva, 1956).

In 1957 at The Ohio State University, Andy took on his first student, Phil Morton, then an undergraduate. Phil tells many interesting tales elsewhere in this volume. In fact, Phil is now retired; still Andy seems to draw no conclusions from this circumstance. At Ohio Andy and visitor Lee Franz completed their work on optical model potentials.[1]

In 1959 Andy made his first trip abroad to the CERN Symposium, to present his work with Carl Nielsen and Keith on the **negative mass instability**,[2] citations of which are ubiquitous in the acclerator literature, papers on gyrotrons, cyclotron emission, circular free-electron lasers, magnetic mirrors, and space-charge effects. This and another paper with Carl on longitudinal space-charge effects were the first of Andy's numerous papers on **collective instabilities**.[3] Andy and family stayed in Ohio until 1959 and then spent one year at LRL.

E. Left the Kids in Denmark

Long about 1961, as an associate professor at Ohio, Andy did a *curious* thing. Rather than stay on for a tenured position, he left for Berkeley and the **Lawrence Radiation Lab**,[4] a place he felt was a kind of mecca, a center of physics.

Leaving Ohio, en route to Berkeley, Andy and family first went to Europe for the summer. That summer of 1961 was spent writing a review article on liquid **helium-3**, at the Niels Bohr Institute, where Andy met Gordon Baym, and Pierre Hohenberg, with whom he continues to interact on **human rights** and **physics planning**. This and a subsequent review article[5] led to a major invited talk the next year---the first talk---at the 8th Low Temperature Conference in London, on Helium-3 work.[6] Leaving the kids in Denmark, Andy and Gladys went to Lake Como where he lectured on liquid helium and water skiing.

---

[1] Pub. #11.
[2] Pub. #17.
[3] Pubs. #12,17,29,30,34,35,38,54,66.
[4] "LRL", the Lawrence Radiation Laboratory, was named after Ernest Orlando Lawrence, inventor of the cyclotron, and the first tenant. It was renamed the Lawrence Berkeley Laboratory in 1971 when local residents realized that Radiation was *BAD* and Berkeley was *GOOD*.
[5] Pub. #2.
[6] Review Article #2 in the c.v.

F. Unpublished Work

At this point in our chronicle, we have a bit of a problem, because a fair bit of Andy's work from the ensuing years wasn't published. There are basically three categories as listed in Table 1:

Table 1. Things Andy Doesn't Publish

(1) Work on things that go "bang" with more than 1MJ/kg (lots of energy---ed.).

(2) Concepts for rapid transport of energy far and above that needed for the next fiscal year.

(3) Romantic correspondence.

We may hear more about categories #1 and #2 in the near future, with the ongoing de-classification program in DOE.[1] In the meantime, we can only guess, when we notice cryptic references to "private communication" or "unpublished" work of AMS.

Here's an example from the *Journal of Mathematical Physics*,[2] in an article entitled "The Hose Instability Dispersion Relation", Steven Weinberg (a high energy theorist---ed.) adds the footnote:

"A. Sessler has shown that the second term in ω_1 is real so that modulating the beam can, at most, affect the scale of the growth rate."

Actually Andy was asked a while back about this reference, by a student grappling with Prof. W's paper. Andy disclaimed any recollection of his work on this, and had no strong feelings for the scale of the growth rate or ω_1, although he has often used this Greek letter in his published work. He did recognize the authors name.

As another example, one finds in *Physical Review D*,[3] in an article by Paul Woodward entitled "Compton Interaction of a

---

[1] "DOE" is the Department of Energy, the folks who provide America's electrical power.
[2] S. Weinberg, J. Math. Phys. **5**, 1371 (1964).
[3] P. Woodward, Phys. Rev. **D 1**, 2731 (1970).

Photon Gas with a Plasma", the citation, of "A.M. Sessler and R. J. Riddell, Jr. (private communication)," following a six-line equation and the remark,

> *"The first order correction terms are in agreement with those obtained earlier by Sessler and Riddell."*

This doesn't tend to bolster one's confidence in the other five lines of the equation, and the first line actually still looks kind of querulous, but Andy at least recognized the author's name and felt comfortable with the possibility of Compton interaction, thought he might even have met Compton.

G. The Lab By the Bay

In Berkeley the Sesslers rented a home for about two years and then built an elegant house. During those years the kids had many pets, did chemistry and electronics experiments, studied music, etc. The Sesslers also did much car-camping when the kids were young, and, when Ruth was old enough, started backpacking and **ski trips**---Andy is an avid skiier, and in later years, in the early 70's, led ski touring groups for the Bay Area Chapter of the Sierra Club, without any serious mishaps.

It was in the first year at LRL that Andy and Jackson and Kelvin Neil, then finishing his Ph.D. thesis under Dave Judd, wrote their now famous works on coherent effects.[1] In 1962 and 1963 Andy took on Victor Wong and Donald Beck who worked in many-body theory. During this time Andy and graduate student John Stack wrote their work on bremstrahlung in dense plasma;[2] John later got his Ph.D. under Geoff Chew. In 1966, after co-authoring with Phil and Kelvin the classic work on wakefields,[3] and with Ernie Courant, the work on transverse resistive-instabilities,[4] Andy had a summer jaunt as **U.S. Advisor to the Panjab University Physics Institute.** This was followed by a

---

[1] Pubs. #20,21.
[2] Pub. #25.
[3] Pub. #34.
[4] Pub. #35.

sabbatical to the **CERN ISR**, about which Eberhard Keil writes in this volume.[1]

At this point in our story, to help us keep track of the chronology, let us recall the *famed Livingston Curve*, depicted in Fig. 2 which correlates highest energy achieved in terrestrial accelerators, versus year. Overlayed we can see some of the highlights in Andy's life.

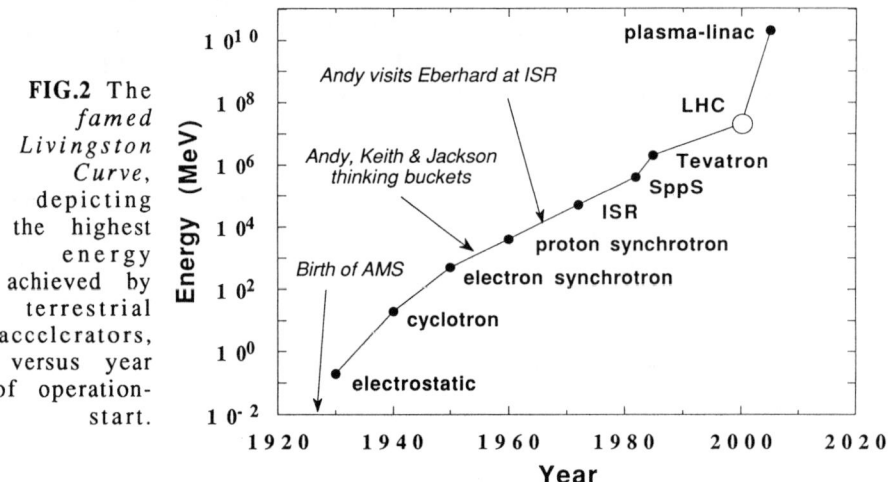

FIG.2 The *famed Livingston Curve,* depicting the highest energy achieved by terrestrial acccelerators, versus year of operation-start.

In 1967 Claudio Pellegrini came to work with Andy, and this collaboration has continued over the years, resulting in their work on the **beam-beam** effect, **bunch lengthening** and much more.[2] Dieter Mohl came in 1969 in the era of the **electron-ring accelerator** (ERA),[3] chronicled elsewhere in this volume. Andy, Jackson and Dieter later performed the original work on the **transverse two-stream** instability.[4] In 1970 Claude Bovet visited from CERN. Andy benefited greatly from the help of Claude, who was a qualified ski instructor and Andy has since

---

[1] "CERN" is the acronym for the European Center for Nuclear Research, "ISR" is the acronym for the *avant-garde* Intersecting Storage Ring. *Avant-garde* then meant $80M.
[2] Pubs.#39,40
[3] Pubs. #45,46,51,52,53,62.
[4] Pub. #64

performed much work in the ski-area with physicists such as Dieter, Herbert Koziol, Ron Ruth, Richard Freeman, Bob Palmer, and Andy Faltens. During this period also Andy pursued his interest in chaos, in particular, chaotic motion resulting from overlapping buckets and the **Chirikov criterion**, with Boris Chirikov and Eberhard.[1]

To have some vignette of these times, consider that in 1971, experiments were beginning at ISR, providing new evidence on the internal structure of the proton, and physicists in the U.S. were devising ways to return to a competitive status---not unlike the situation today. In the summer of 1971 SLAC held a two-week summer study on a Recirculating Linear Accelerator (RLA) upgrade for SLAC, which Andy and sixty other physicists attended including Dieter and Claudio. One year later Wolfgang K. H. Panfosky would be testifying to Congress on behalf of this upgrade, SPEAR and numerous other projects within the purview of the AEC. With him Robert B. Wilson would be testifying on behalf of the 200 BEV NAL at Batavia then getting its first beam, sort of.[2] DESY was two years away from completion. In this atmosphere a practical proposal for colliding positrons or electrons with protons, the **PEP** concept, was introduced by Claudio, Dieter, Andy, Burton Richter, John Rees, and Mel Schwartz. Appearing more attractive than the RLA, PEP was eventually funded, minus the protons that were deemed too expensive.[3]

In 1971 LBL consisted of 1,025 personnel including 99 physicists, 87 graduate students, and 1 Sessler, roughly equal in manpower to BNL or SLAC, and roughly 10% of the totals in all the National Accelerator Labs.[4] The big machine at LBL was the Bevatron, which had been operating in some form since 1954 (original construction cost $10M, upgraded in 1963). The *next-generation* accelerator of that time was the ERA, championed by

---

[1] Pubs. #7,50,69.

[2] This testimony by Dr. Wilson includes the now famous account of NAL's exploding magnets, lunch-bags in the beamline and the Amazing Beam-Ferret "Felicia".

[3] Pub. #60,66. "PEP" started out as the acronym for the Positron-Electron-Proton project, and after the Protons were cancelled, became the Positron-Electron-Project.

[4] The source for these numbers and an excellent reference on accelerator activity during this time is the AEC Authorizing Legislation, Fiscal Year 1973, Hearings before the Subcommittee on Research, Development, and Radiation, Joint Committee on Atomic Energy, 92nd Congress.

Andy, Jackson, Denis Keefe, Andy Faltens, and others. ERA had been under development for three years, with two experiments completed at the ASTRON accelerator at Livermore (in '68 and '69). The LBL ERA staffing level for '71 was 48 man-years, with a projected cut of 14 man-years for '72.[1] A 100 GeV ERA pilot-project had been proposed. The Congressional Record shows that on 7 January, 1971 at 1:15PM one Andrew Sessler was giving a 45-minute talk on "Status of ERA Theory and Concepts" to the HEPAP Subpanel on Accelerator Technology meeting at LBL.

This same Andy became the **Director** of LBL two years later, as described by Earl Hyde elsewhere in this volume, including the story about the deer and the policeman, and how Andy decided to lump Accelerator and Fusion Research into one Division (AFRD), leaving Physics Division little to do but particle phyics. These were the days when Luis Alvarez would stop by every day or two to discuss progress on his theory of extinction at the KT boundary, at that time, a kind of personal pursuit of his, supported by the Director's Fund.

As Director Andy continued his research activities, taking on Bill Sharp as a post-doc in 1974, and in 1975, taking on his next student, Sid Putnam, who had been working at Physics International for some time on collective-effect acceleration methods (ERA). During the same period, Andy took on a new grad student Paul Channell and they together performed the seminal work on **turbulence in beams**, in connection with bunch lengthening and widening.[2] It is during this period that the mysterious episode of the "burning car" occured on Route 280, on the way to an Advisory Committee meeting for PEP. Andy gave up pipe-smoking shortly thereafter.

During the late 70's Andy's physical condition deteriorated and he developed a kind of general malaise affecting his spirit. In 1980 he resigned the Directorship. The reason for this malaise soon became apparent.

H. The Operation

In the Spring of 1980 the Doctor came to Andy and told him he had developed a menangioma of the brain stem, a tumor, and the chances of surviving an operation were pretty slim. Andy

---

[1] ERA work was also being pursued by Sarantsev at Dubna, and in Munich and Karlsruhe. Total staffing world-wide was about 200 people.
[2] Pub. #69.

delayed the operation by a week as he had been scheduled to give a talk at Don Kerst's retirement. He went to Don's affair, gave his talk, skipped the reception, and took a plane back from Wisconsin to the West Coast. Gladys met him at the San Francisco Airport and took him directly to the Redwood City Kaiser (convenient to the Airport). The operation started first thing in the morning and lasted eight hours.

And it was a success; however there was some loss of balance and hearing as a result of which Andy had to relearn walking, and has ever since had to concentrate extra-hard on listening. He also gave up bike-riding, at least for the time being. There was a day when Andy rode a great deal, usually a business day---a one-day 100-mile ride not being unusual, and his "personal best" a "100 miles-10,000 feet of climb" in one day. Since that time Andy has satisfied himself with jogging about fifteen miles a week---much easier while jogging to carry on a good physics conversation, quizz students, and preserve a healthy tan. He has been clocked at about 8 minutes a mile, 9-10 whenever the "new idea" of the day was revealing serious physical flaws.

I. Great Strides All Morning

In the 1980's, freed from the duties of Director Andy's career really started to pick up. Jin-Soo Kim, now at UC San Diego, was his next student, together with Jonathan Wurtele, now at MIT.[1] It was during this era that the concept of a TeV electron-positron linac was coming to be seen as the *next generation* accelerator, and to this end Andy proposed the **two-beam accelerator.**[2] In 1983 Andy took on a new student Efrem Sternbach and began his work on the **Electron Laser Facility** experiments at Lawrence Livermore National Laboratory (LLNL) as described elsewhere in this volume. Simultaneously, Andy and colleagues proposed the concept of **optical guiding.**[3]

In 1985 Andy visited **KEK**[4] for two months under a fellowship from the Japan Society for the Promotion of Science, at

---

[1] "MIT" is the Massachusetts Institute of Technology, Cambridge, Massachusetts, trains engineers, recently started an English Department.
[2] Pub. # 74,103.
[3] Pub. #80,83,98,140.
[4] "KEK" is Ko-Energie Butsuri Gaku Kenkujo, The National Laboratory for High Energy Physics, in the Tsukuba metropolitan area, Japan.

a time when TRISTAN was still under construction. People at KEK are still talking about Andy-sensei effortlessly running laps around TRISTAN in the hot August noon, querying his fellow joggers, and offering detailed comments on the TRISTAN layout and prospects for rice-farming.

During the period 1985-86 Andy served on the **A P S Directed Energy Study Committee**, now famed for coming to the wrong conclusions on SDI.[1] During 1987-88 Andy was Vice-Chairman (later, Chairman) of the **Federation of American Scientists**. In this time the APS topical group "Physics of Beams" was coming into its own. For several years Mel Month had been championing a **"Division of Beam Physics"**, and he had prepared a White Paper. Andy and Mel and others made a forceful case to the APS Council, and the Division of Beam Physics was born, with Andy as Chairman. Actually Andy's professional life has been marked by such "public service", service for the DOE, the NSF, AUI, BNL on numerous committees, editor of Nuclear Instruments and Methods for twenty years, refereeing of papers and proposals for numerous journals and sponsors, advising on personnel actions for many universities and labs---almost a full-time job.

To have some picture of Andy's *modus operandi* in these and later years, let's look at a typical day. The crew in these days included: Yehuda Goren and Don Hopkins working in nearby offices on microwave problems, (Don fresh from his 2GW victory with the ELF experiment); a new grad student David Whittum; Gil Travish, an undergraduate at that time, working on beam break-up; Efrem Sternbach, working on FEL's; and Jeff Tennyson working on beam-beam interactions, usually accompanied by his dog Cy ("The Beam Dog"). A typical day would start with Andy coming in around 8 AM and making coffee. His first words were usually "Well I've been thinking about the [new idea]...and you know it really just doesn't seem right what you said..." As the coffee bubbled down, and the always just-upgraded computer system began to hang, Andy would produce calculations written in a kind of languid cursive script revealing a much simpler way to understand the new idea. Every ten minutes or so this scene would be repeated in some fashion with each member of the crew.

---

[1] E. Teller, private communication. SDI is the acronym for a movie by George Lucas.

By 9 AM Andy was free to water the plants, take out the garbage, referee a paper, return phone calls, answer email and prepare abstracts for upcoming conferences. Sounds from the other offices were quiet mumbles, and the tone of Macintosh computers being re-booted. Sometime in the morning Bill Fawley or Simon Yu or Bill Barletta or Bill Sharp would stop by to correct major errors in Andy's thinking. As noon approached his crew would start to filter back into Andy's office, claiming to have vindicated their original thinking (momentum conserved, Liouville happy, Maxwell satisfied), or disowning their original thinking but having uncovered new problems (emittance too low, current too high, non-linearity kills it, linearity allows instability to kill it, idea is not new---worked out by Russians in '52). In a timely fashion, Andy would begin to change into his jogging shorts, to go running.

After his run and a small lunch at his desk, Andy would return to the foray, striding about, cheerfully motivating the troops, repelling assaults with order of magnitude estimates and returning to his office to prop his feet on the desk and considering deeply New Things as yet not identified to his crew. In the evening usually about 5PM, declaring that "great strides" had been made, "possibly in circles", but still great strides, Andy would go out carousing, always claimed he was going to the **opera**, or **recorder group** practice. The lights in the offices would burn into the night, numbers crunching, Macintosh computers rebooting---could this guy be right after all?

In 1988 Salvatore Solomeno visited for a year to work on FEL's and in 1989 John Dawson came from UCLA to spend a sabbatical year. This in the midst of what we might call Andy's **beam-plasma** period, including such works as the **adiabatic focuser**,[1] **plasma suppression of beamstrahlung**[2] and the **ion-channel laser**.[3] During this period also Andy and John invented a new method, now patented, of sequencing **DNA**,[4] finally offering Andy's heirs' some hope of actually benefitting from his work. This was closely followed by the invention of microwave **beam conditioning**[5] for FEL's, with Li-Hua Yu and a

---

[1] Pub. #131.
[2] Pub. #112.
[3] Pub. #133.
[4] Pub. #125.
[5] Pub. #141.

former grad student. Throughout this period Andy, Bill Barletta, and others were working on sundry beam-plasma radiation techniques.

In recent years Andy has continued to pursue his interest in FEL's, notably the **standing-wave free-electron laser** and other **FEL oscillators**, working with post-docs Srinivas Krishnagopal, Govindan Rangarajan, Chang-Biao Wang and Hai Li, and others of the **Center for Beam Physics**, Ming-Xie, Yong-Ho Chin and Kwang-Je Kim. In addition, Andy continues to work on the **relativistic klystron** two-beam accelerator proposed with Simon Yu, and later built at LLNL by Glen Westenskow and Tim Houck. Giulia Fiorentini visited in 1991 to work on that. More recently, Marieke de Loos and Sebastiaan van der Geer are visiting from FOM in Holland to pursue their interest in radiation sources.

Andy has also supervised numerous undergraduates over the years, including Gil Travish, now at UCLA, Jennifer Eden, Nicholas Goffeney and Richa Govil who is now his graduate student and is working on **plama-lens** experiments at the ALS[1] and other projects. Andy's latest studies with Hiromi Okamoto, Jie Wei and Xiao-Ping Li include the fundamental theory of **crystalline beams**, and **beam cooling** techniques. Andy has also over the last five years or so been pursuing **frequency shifting** in plasma with Chan Joshi and Warren Mori at UCLA, and this has taken him into astrophysics, with his most recent visitor Vinod Krishnan. At the same time, Andy continues to pursue beam-plasma, and *next-generation* plasma acceleration studies with Wim Leemans and Swapan Chattopadhyay at the Center for Beam Physics, and collaborators Pisin Chen at SLAC and Tom Katsouleas and his students at USC. Experiments are planned shortly for the ALS.

J. Works in Progress

Andy's son Jon is now a Professor of Chemistry and his son Dan is a Medical Doctor. In fact, since 1991, Andy has been collaborating with Dan on medical research in **pre-operative thermal insulation**.[2] There are three grand-children: **Zachary**, age 6, **Ethan**, age 4.5 and **Jordan** age 3. Jordan began skiing at

---

[1] "ALS" stands for Advanced Light Source, an accelerator for producing *GOOD* radiation.
[2] Pubs. #138,149,156.

age two and currently is pursuing advanced ski-work, including a stuffed animal in each arm---"to let them enjoy the experience," he tells his mother, who has other theories. At Christmas-time Andy was playing flute to Zachary's piano pieces. Zachary had been studying for a year, Andy, of course, for many years, and finally seemed to have found his level. Ethan likes to play doctor on Andy, giving him pills that turn him into a lion, a wolf, etc. Andy gave him a "grown up" pill and found Ethan in the kitchen later making coffee.

Andy also enjoys writing Letters to the kids and so far none have been rejected. Letters about anything, like when the kids got a hot-tub, Andy mentioned to Zachary that he had noticed a shark living on the bottom. A good year later Andy was in the hot-tub with Zachary when Andy felt something *bite his toe*. Andy screamed, but Zachary explained that it was only the shark. Andy also notified the kids once when a cookie factory broke in Oakland and was turning out cookies *like mad* (chocolate, etc.)---and would they please come quick and help eat the cookies as all the local kids were too full.

This year Andy has arranged ski vacations with Jon's family in Utah, and with Ruth's family in Montana. And he spent a week at Thanksgiving with Jordan, *et al.* and a week at Christmas with Zachary, Ethan, *et al.* But Andy travels extensively, even when he's not imposing on his kids. Not just professional travel (9 times to the Soviet Union, N times to Europe, where N is large, several trips to Japan), but recreational travel to Israel, Egypt, Tunis, Kenya, Peru, Nepal, Patagonia, Costa Rica, The Lost World, in Venezuela...Andy has hiked to the Annapurna Sanctuary (13,500 feet), and across a 16,500 foot pass in the Andes once when he lost his map. One wonders, how he could do it all on Lab time!

III. Prospects for the Future
"How can I, those students standing there, my attention fix, on Livermore or SLAC or RadLab politics..."

Ordinarily a story can be ended without great fanfare, but this has been a very long story and, in any case, we are physicists and always long to know the Prospects. To make some predictions, let us recall Fig.1, a histogram of Andy's publications by year. In order to extrapolate from such data, we adopt the most conservative approach possible, that used by economists in making crucial predictions about the national economy, *i.e.*, an exponential fit.

The *cumulative* number of publications is shown in Fig. 3, with an extremely good exponential fit overlayed. Now that mandatory retirement is defunct, and relying on standard actuarial tables, we can predict 2000 publications by the year, 2010. Approximately 300 of these will be unrefereed contributions to particle accelerator conferences and workshops, 400 will be published in Physical Review E, 100 in Physical Review Letters, and the rest will be book-reviews, with a margin of error of 10%.

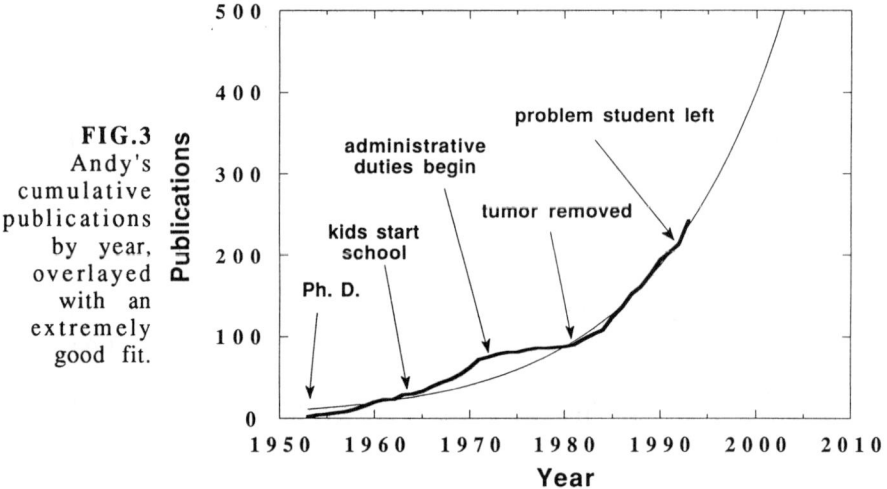

FIG.3 Andy's cumulative publications by year, overlayed with an extremely good fit.

A similar analysis can be applied to Andy's student-body count. In Fig. 4 we see the cumulative distribution, with a careful extrapolation over the next thirty years. According to this estimate, by the year 2030, Andy's students will encompass almost 10% of the population of Los Alamos, New Mexico.

There is a larger significance to Andy's curves, however. Recall Fig. 2, the *famed Livingston Curve*, which correlates highest energy achieved in terrestrial accelerators, versus year. The inference commonly drawn from this curve is that there is some correlation between energy and year. Actually however, the true cause and effect relation is depicted in Fig. 5, a plot of energy versus Andy's cumulative publication count. This *Sessler Curve* yields a χ^2 error a factor of ten lower than the conventional *famed Livingston Curve*.

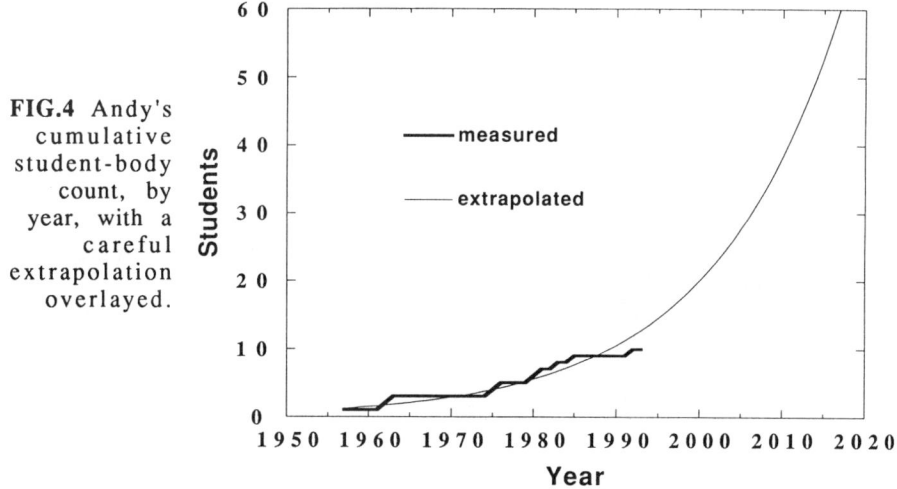

FIG.4 Andy's cumulative student-body count, by year, with a careful extrapolation overlayed.

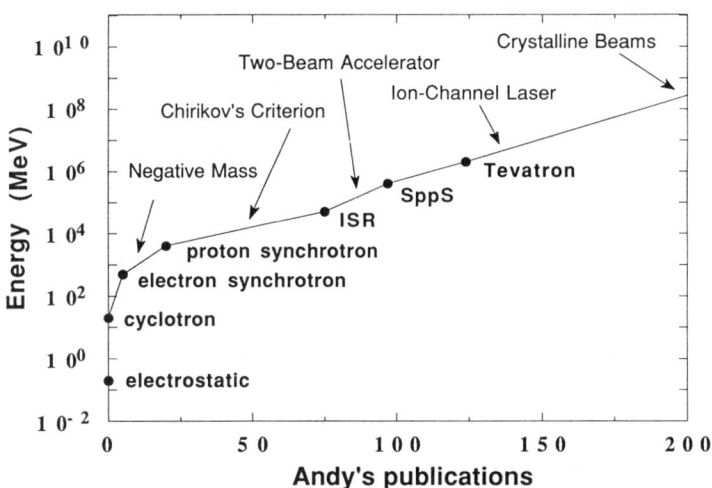

FIG.5 The *Sessler Curve*, depicting highest energy achieved versus Andy's cumulative publication count. Advanced statistical procedures reveal a factor of ten higher correlation of energy with Andy's work, than with time as conventionally measured.

As we gaze at these plots there is one clear conclusion: Andy will need **more funding**! Enjoy this volume! It was sixty-five years in the making. See you at the next Sessler Symposium in 2003, with more stories to tell!

ANDY SESSLER: A PHYSICIST FOR ALL SEASONS
(Sessler Symposium on the Physics of Beams, in honor of Andy's 65th birthday)
December 5, 1993

Morris Pripstein
Lawrence Berkeley Laboratory
Berkeley, CA 94720

Most of you are aware of Andy's long and distinguished career as a theoretical physicist and as a former director of LBL at a critical stage in the lab's history. However, much less known and yet as important, if not more so, is Andy's truly major contribution on behalf of the human rights of fellow scientists around the world. Many lives have been affected by Andy's efforts. To know what Andy has done in this area is to gain a special insight into this remarkable moral giant.

If one looks at Andy's curriculum vitae, one sees that he was chairman of the American Physical Society's human rights committee (CIFS), president of the Federation of American Scientists and a co-founder of Scientists for Sakharov, Orlov and Sharansky (SOS). The list, by itself, does not begin to convey his accomplishments, other than to indicate that he was probably active. I had both the privilege and pleasure to be one of the founding group of SOS, and I would like to describe some of Andy's activities in that group because, in many ways, they reflect and summarize his unique role as a human rights activist.

When Andy was director of LBL he had adopted an open-door office policy in his interactions with the lab staff, which was great for us if not for Andy. Encouraged by this policy, many of the scientists, myself included, often went in to see Andy, usually to complain about some lab issue of the moment. To his credit, Andy never lost his equanimity, even when unduly provoked from time-to-time by some group leaders and ex-group leaders who shall remain nameless. As a result, we felt no inhibitions in discussing our concerns with Andy. One day in late May 1978, when Denis Keefe of LBL I met with Andy about a lab problem, we expressed our deep concern about the worsening plight of our colleagues in the Soviet Union. This was evident from the recent harsh prison sentence of accelerator physicist and human rights activist Yuri Orlov, and the dire threats of possible death penalty for "treason" in the impending trial of computer scientist Anatoly Sharansky for his human rights activities.

Andy, ever the activist, responded by saying that we must resort to new tactics to confront this heightened repression by the Soviet authorities. Traditional expressions of concern and pleas for moderation were apparently no longer of much use. Andy encouraged us to organize a small group of like-minded scientists at LBL, including himself, to develop new ways to help our colleagues. This led to the formation of Scientists for Orlov and Sharansky (SOS) which later evolved into Scientists for Sakharov, Orlov and Sharansky, keeping the same acronym, SOS.

As we began to plan a course of action, it became evident that Andy would be increasingly in the public eye. Many of us worried that this may adversely affect Andy's relationship as lab director with the Department of Energy. Andy had no such fears. He was adamant that his human rights activities were part of his responsibilities as an individual citizen and in no way conflicted with his duties as lab director. Freed from this restraint, we then proceeded to take advantage of Andy, by having him speak at countless public rallies and press conferences, in addition to his extensive letter-writing campaigns.

The first major public initiative occurred in July 1978, during the Sharansky trial in Moscow. We invited Sharansky's wife, Avital, who was then living in Israel, to come to the U.S. and meet with scientists around the country to galvanize support for her husband within the scientific community. This was financially problematic as we had no money whatsoever to do this; however, Andy blithely suggested we charge her trip expenses to our personal credit cards in the hope that we would eventually be reimbursed by future donations from our colleagues. Fortunately for us, Andy's prediction was borne out.

Locally, we organized a highly successful rally on the Berkeley campus, which included the folksinger Joan Baez and Andy, as well as Mrs. Sharansky. There were 5,000 people in attendance, making it the largest rally there since the end of the Vietnam war. Even more remarkable was that this large turnout occurred during the summer when most students were away. Andy's speech in defense of human rights in general, and Sharansky in particular, was especially eloquent and poignant, bringing tears to the eyes of many in the audience. It also brought us more volunteers for future efforts.

Spurred on by the enthusiastic response to this first venture, we organized a moratorium on scientific exchange between individual American scientists and the Soviet Union, as a protest of the blatant mistreatment of our Soviet colleagues by the Soviet authorities. In an action unprecedented in science, except for the period of World War II, more than 2,400 American scientists signed the moratorium pledges. Many of the signatories had previously been in the vanguard of promoting exchange between the two scientific communities. In the Washington press conference of February 1979, to announce the results of this moratorium campaign, Andy prepared an eloquent statement, summarizing his stand on human rights and his view of the role of SOS. Since it is quintessential Sessler, I should like to quote it here:

Statement Concerning SOS, by Andrew M. Sessler (February 26, 1979)

> I have, as long as I can remember, been concerned about human rights for, in the spirit and words of Thomas Jefferson, "I have sworn, upon the alter of God, eternal hostility against every form of tyranny over the mind of Man."

> I became involved with the plight of Yuri Orlov when he was first arrested about two years ago. I knew him as a scientist of

considerable distinction; in fact I knew him well for we had met on four different occasions. I was also familiar with his activities as a founder of the Committees to monitor the Helsinki Accords. I considered his arrest an outrage and immediately took the initiative in organizing a telegram of protest which was signed by 200 physicists and sent out within three days of Orlov's arrest.

Subsequently, I partook in the organization of SOS, primarily because I believe that scientists and engineers--acting as individuals--can have considerable influence upon the Soviet Union. The direct action, as evidenced by the many who have signed the SOS petitions, will have an effect, for clearly the Soviets need international scientific cooperation to achieve and maintain first-rate science and technology. Thus by our activities, we expect to eventually achieve the release of Orlov and Sharansky and, more generally, modify the Soviet Government attitudes and actions against those Soviet citizens who are doing no more than exercising their God-given human rights.

Reaction from the Soviet Union to our initiative was swift and strong. Within weeks, there appeared a smuggled letter from Andrei Sakharov and the Soviet mathematician, Naum Meiman, written on behalf of many dissidents, strongly praising our actions. At the same time, we were denounced in the Soviet media, first on Moscow radio news and then in a very long article in Pravda. Andy was particularly amused by the radio denunciation because the well-known Russian commentator, Valentin Zorin, concluded that the "innocent" American scientists were led astray by evil leaders such as Senator Robert Dole. Aside from the fact that non-scientists were in no way involved, the political leader whose views were perhaps farthest from Andy's was Senator Dole.

Conditions in the Soviet Union continued to deteriorate. The Soviet invasion of Afghanistan was accompanied by further repression of human rights within the Soviet Union, culminating in the exile of Andrei Sakharov to Gorky in January 1980. With Andy prodding us to respond, we in SOS escalated our activities and organized a worldwide moratorium of science exchange with the Soviet Union. This resulted in 7,900 scientists from 44 countries, to commit themselves to this course of action as their most forceful way to register their support of their beleaguered colleagues. A number of those signing up came as a result of Andy's personal letter-writing campaign to augment our more public efforts. Amidst much publicity, we sent off the list of signatories to President Brezhnev and to Soviet Academy of Sciences President Alexandrov.

Soon after, Andy's sense of humor led us to an unusual encounter. We learned that a Soviet Science Attaché who was apparently also a member of the KGB, wanted very much to get a copy of this list, which was now in the public domain but with very limited circulation. So, Andy in his inimitable way, suggested that we accede to this request but with the proviso that the Attaché meet us for lunch (on our dime, no less!) so that we could discuss some of the issues, in a pleasant setting, while giving him the information he desired. The

Attaché warily accepted the invitation and joined Sessler, Denis Keefe and me, of SOS, for what turned out to be a very long lunch. Andy explained at great length to this fellow, who was trapped at the table, the depth of anger felt by Western scientists over the mistreatment of their Soviet colleagues. His relief upon being able to leave finally at the end of the meal was palpable. It is amusing to envision him reporting his impressions of Andy to his superiors.

During this period, we also adopted the strategy of publicly protesting visits to the U.S. of prominent Soviet scientists who had led campaigns of vilification against Andrei Sakharov and other dissident Soviet colleagues. On several occasions we established picket lines on the streets outside the meetings where the Soviet scientists were to appear. In each case, the media were there to report our actions and Andy was quoted extensively, in language frank if not diplomatic. For example, he referred to one Soviet Nobel Laureate as "a great scientist but a lousy human being" and to another as "notorious as a virulent antagonist of Sakharov". Phrases not quite up to the elegance of Andy's SOS statement above but certainly clear and unambiguous in conveying his opinions to the public.

Andy's unflinching courage on behalf of human rights was best illustrated in another major SOS initiative for the Sakharovs. During the time of Sakharov's exile in Gorky, his wife, Elena Bonner, had suffered severe heart problems and required by-pass surgery in the West. The Soviet authorities consistently refused to allow her to leave the country even temporarily for this medical treatment, alleging that she would use the opportunity to vilify and embarrass the Soviet Union. In 1984, after many futile pleas for her release, we decided to challenge the literal excuse of the Soviet authorities, by organizing a hostage exchange for the release of Ms Bonner -- Western scientists to go to the Soviet Union and stay there while Bonner was in the U.S. for medical treatment. Their stay there would be as "good-faith witnesses" or hostages to guarantee that Bonner's visit to the U.S. would be for purely medical purposes. Despite the serious risk involved, Andy was one of the first to volunteer as a hostage. By contrast, many scientists were understandably reluctant to participate for fear of the possible consequences. Nonetheless, thanks to Andy's example among others, we got fifty-five prominent scientists from thirteen countries to join this initiative, which received world-wide media attention -- but no immediate response from the Soviet authorities.

With all of Andy's varied activity, including constant letter-writing on behalf of threatened colleagues, there was a happy ending. Approximately a year later, Elena Bonner was allowed out of the Soviet Union for medical treatment in the U.S., and a few months later, in 1986, our colleagues Orlov and Sharansky were released from labor camp and left the Soviet Union. This was followed soon after by Sakharov's release from exile in Gorky. One of Andy's great pleasures, I believe, was to subsequently greet all these people personally in Berkeley and to receive the thanks of many others who he has helped.

To conclude, I want to say to Andy: Your human-rights efforts have been an inspiration to us all. You are truly a physicist for all seasons.

Curriculum Vitae

ANDREW M. SESSLER
Physicist

Lawrence Berkeley Laboratory
University of California
1 Cyclotron Road
Berkeley, California 94702, USA
(510) 486-4992
e-mail: TBALBL@lbl.gov

EDUCATION

A.B. Cum laude, Harvard University - 1949, Mathematics

M.A. (1951) and Ph.D. (1953) - Columbia University Theoretical Physics

PROFESSIONAL EXPERIENCE

Assistant in Physics (1949-52)

National Science Foundation Predoctoral Fellow (1952-53)

National Science Foundation Postdoctoral Fellow (1953-54)
with Prof. H.A. Bethe, Cornell University

Assistant Professor, The Ohio State University (1954)

Associate Professor, The Ohio State University (1960)

On leave to Midwestern Universities Research (1955-56)

Visiting Physicist, Lawrence Radiation Laboratory (1959-60)

Visiting Physicist, Niels Bohr Institute in Copenhagen (Summer 1961)

Visiting Physicist, Intersecting Storage Ring Division of CERN, Geneva, Switzerland (1966-67)

Japan Society for the Promotion of Science Fellowship at KEK (Tsukuba, Japan) (Fall 1985)

© 1996 American Institute of Physics

Lawrence Berkeley Laboratory

    Theoretical Physics: Research on many body problems, accelerator design problems, and plasma physics (1961-73)

    Energy and Environment: Participated in development of program (1971-73)

    Director: Lawrence Berkeley Laboratory (1973-1980)

    Senior Scientist: Research on plasma physics and accelerator physics (1980-)

PROFESSIONAL ACTIVITIES

    U.S. Advisor to the Physics Institute at Panjab University, Chandigarh; U.S. -India Cooperative Program for the Improvement of Science Education in India (Summer 1966)

    Member, High Energy Physics Advisory Panel to the U.S. Atomic Energy Commission (1969-72)

    Correspondent to Comments on Modern Physics: Part Λ (1969-71)

    Member, Editorial Board of Nuclear Instruments and Methods (1969-....)

    Member, Lawrence Hall of Science Advisory Committee (1974-78)

    Chairman, Stanford Synchrotron Radiation Project (now SSRL) Science Policy Board (1974-77)

    Chairman, EPRI Advance Fuels Advisory Committee (1978-81)

    Chairman, BNL External Advisory Committee on Isabelle (1980-82)

    Member, Princeton University Review Committee for the Plasma Physics Laboratory (1981-85)

    Chairman, American Physical Society Committee on International Freedom of Scientists (1982)

    AAAS Nominating Committee (1984-87)

    Member, Federation of American Scientists Council (1979-82, 1985-88)

    Chairman, ICFA Panel on Novel Accelerators (1984-87)

    Member, Committee of Concerned Scientists Council (1984-87)

    Chairman, NSF Gravity Wave Observatory Panel (1986, 1987, 1993)

    Member, American Physical Society Study of Directed Energy Weapons (1985-86)

Chairman (1988) Panel on Public Affairs, American Physical Society
    Vice Chairman (1987)

Chairman, Vanderbilt University FEL Panel (1987)

Chairman, Federation of American Scientists (1988- 1991)
    Vice Chairman (1987-88)

Member, American Physical Society Division of Physics of Beams (Chairman 1990)

Member, Board of Directors, Associated Universities, Inc. (1991-   )

Member, Stanford Synchrotron Radiation Laboratory Science Policy Board (1991-1992)

Chairman, American Physical Society Committee on the Applications of Physics (1993)

Member, Honorary Advisory Board, Institute for Advanced Physics Studies, LaJolla International School of Physics (1991-   )

Member, Superconducting Super Collider Scientific Policy Committee (1991-1993)

Member, American Physical Society Council, Divisional Councilor for DPB (1994-1997)

PROFESSIONAL MEMBERSHIPS AND AWARDS

Fellow, American Physical Society

Fellow, American Association for the Advancement of Science

Member, Sigma Xi

Member, New York Academy of Sciences

Senior Member, Institute of Electrical and Electronics Engineers

Finalist, Westinghouse Science Talent Search (1945)

Recipient, E.O. Lawrence Award by Atomic Energy Commission (1970)

Recipient, US Particle Accelerator School Prize (1988)

Member, National Academy of Sciences (elected 1990)

Leland J. Haworth Distinguished Scientist, Brookhaven National Laboratory (1991-1992)

Nicholson Medal For Humanitarian Service, The American Physical Society (1994)

RESEARCH PAPERS

1. A.M. Sessler and H.M. Foley, "Spin-Spin Interaction of Electrons and the Ionization Energy of Helium", Phys. Rev. 92, 1321 (1953).

2. A.M. Sessler and H.M. Foley, "The Relativistic Correction to the Ground-State Energy of Helium", Phys. Rev. 92, 1321 (1953).

3. A.M. Sessler and H.M. Foley, "Statistical Atom with Angular Momentum", Phys. Rev. 96, 366 (1954).

4. A.M. Sessler, "Mesonic Corrections to the Quadrupole Moment of the Deuteron", Phys. Rev. 96, 793 (1954).

5. A.M. Sessler and H.M. Foley, "Hyperfine Structure of He 3+ and He 3", Phys. Rev. 98, 6 (1955).

6. D.W. Kerst, et al., "Attainment of Very High Energy by Means of Intersecting Beams of Particles", Phys. Rev. 102, 590 (1956).

7. K.R. Symon and A.M. Sessler, "Methods of Radio Frequency Acceleration in Fixed Field Accelerators", Proceedings of the CERN Symposium Vol. 1, 44 (1956).

8. D.W. Kerst, et al., "Operation of a Spiral Sector Fixed Field Alternating Gradient Accelerator", Rev. Sci. Inst. 28, 970 (1957).

9. A.M. Sessler and R.L. Mills, "Nucleon Size Contributions to the Hyperfine Structure of Hydrogen and Helium", Phys. Rev. 110, 1453 (1958).

10. A.M. Sessler and H.M. Foley, "Hyperfine Structure of Deuterium and Nucleon-Nucleon Spin-Orbit Potentials", Phys. Rev. 110, 995 (1958).

11. L.M. Frantz, et al., "Exclusion Principle and Phenomenological Optical-Model Potentials", Phys. Rev. Letts. 1, 340 (1958).

12. C.E. Nielsen and A.M. Sessler, "Longitudinal Space Charge Effects in Particle Accelerators", Rev. Sci. Inst. 30, 80 (1959).

13. L.N. Cooper, R.L. Mills and A.M. Sessler, "Possible Superfluidity of a System of Strongly Interacting Fermions", Phys. Rev. 114, 1377 (1959).

14. R.L. Mills, A.M. Sessler, S.A. Moszkowski, and D.G. Shankland, "Superfluidity of Nuclear Matter", Phys. Rev. 3, 381 (1959).

15. V.J. Emery and A.M. Sessler, "Energy Gap in Nuclear Matter", Phys. Rev. 119, 248 (1960).

16. V.J. Emery and A.M. Sessler, "Possible Phase Transition in Liquid He3", Phys. Rev. 119, 43 (1960).

17. C.E. Nielsen, A.M. Sessler and K.R. Symon, "Longitudinal Instabilities in Intense Relativistic Beams", Proc. of CERN Symposium High Energy Accelerators (1959) p. 239.

18. S.A. Moszkowski and A.M. Sessler, "Hole-Hole Interactions and the Properties of Nuclear Matter", Nuclear Physics 18, 669 (1960).

19. A.E. Glassgold and A.M. Sessler, "Flow Properties of Superfluid System of Fermions", Il Nuovo Cimento Series 10, 19, 723 (1961).

20. V. K. Neil and A.M. Sessler, "Coherent Electromagnetic Effects in High-Current Particle Accelerators: I. Interaction of a Particle Beam with an Externally Driven Radio-Frequency Cavity", Rev. Sci. Inst. 32, 256 (1961).

21. L.J. Laslett, V. K. Neil, and A.M. Sessler, "Coherent Electromagnetic Effects in High-Current Particle Accelerators: III. Electromagnetic Coupling Instabilities in a Coasting Beam", Rev. Sci. Inst. 32, 276 (1961).

22. D.W. Kerst, et al., "Electron Model of a Spiral Sector Accelerator", Rev. Sci. Inst. 31, 1076 (1960).

23. A.M. Sessler, "Possible Low-Temperature Superfluid Phase of Liquid He3", Proceedings of the Second Symposium on Liquid and Solid He3, Edited by John G. Daunt (Ohio State Press), 81-102 Aug. 23-25, 1960.

24. L.J. Laslett and A.M. Sessler, "Coupling Resonances in Spiral Sector Accelerators", Rev. Sci. Inst. 32, 1235 (1961).

25. J. D. Stack and A.M. Sessler, "Bremsstrahlung in a Dense Plasma," The Physics of Fluids 6, 1193 (1963).

26. G. Baym and A.M. Sessler, "Perturbation -- Theory Rules for Computing the Self-Energy Operator in Quantum Statistical Mechanics", Phys. Rev. 131, 2345 (1963).

27. H.P. Kelly and A.M. Sessler, "Correlation Effects in Many-Fermion Systems: Multiple-Particle Excitation Expansion", Phys. Rev. Lttr. 132, 2091 (1963).

28. F.T. Cole et al., "MURA 50-MeV Electron Accelerator -- Design Study and Choice of Operating Point II", Rev. of Sci. Inst. 35, 1398 (1964).

29. V.K. Neil and A.M. Sessler, "Longitudinal Resistive Instabilities of Intense Coasting Beams in Particle Accelerators", Rev. Sci. Inst. 36, 429 (1965).

30. L.J. Laslett, V.K. Neil and A.M. Sessler, "Transverse Resistive Instabilities of Intense Coasting Beams in Particle Accelerators", Rev. Sci. Inst. 36, 436 (1965).

31. A.M. Sessler, "SLAC Storage Ring Summer Study Summary Report Contributions", -Nos. 2, 3, 4, 6, 9, 10 (1965).

32. A.M. Sessler, "Beta-Ray Spectrometer With Reduced Spherical Aberration", Nucl. Inst. and Methods 23, 165 (1963).

33. L.J. Lasett and A. M. Sessler, "Rotation of Mercury: Theoretical Analysis of the Dynamics of a Rigid Ellipsoidal Planet:", Science 151, 1384 (1966).

34. P.L. Morton, V.K. Neil and A.M. Sessler, "Wake Fields of a Pulse of Charge Moving in a Highly Conducting Pipe of Circular Cross Section", Journal of Applied Physics 37, 3875 (1966).

35. E. D. Courant and A.M. Sessler, "Transverse Coherent Resistive Instabilities of Azimuthally Bunched Beams in Particle Accelerators", Rev. of Sci. Instr. 37, 1579 (1966).

36. K. Bergkvist and A.M. Sessler, "A High Resolution, High Luminosity Beta-Ray Spectrometer Design Employing Azimuthally Varying Magnetic Fields", Nuclear Instr. and Methods 46, 317 (1967).

37. D.E. Beck and A.M. Sessler, "Properties of Liquid Helium-Three in the Two-Body Correlation Approximation I", Phys. Rev. Lttr. 146, 161 (1966).

38. A.M. Sessler and V. Vaccaro, "Longitudinal Instabilities of Azimuthally Uniform Beams in Circular Vacuum Chambers With Walls of Arbitrary Electrical Properties", CERN Report, Geneva, 1967, CERN 67-2.

39. C. Pellegrini and A.M. Sessler, "Transverse Coherent Beam Phenomena in Colliding Beam Devices", Proceedings of the VIth International Conference on High Energy Accelerators (Cambridge, 1967), p.135.

40. C. Pellegrini and A.M. Sessler, "Curvature Effects and the Shape of Bunches in Electron Storage Rings", Nuovo Cimento (1968), Series X, 53B, 198.

41. E.D. Courant, E. Keil and A.M. Sessler, "Improvement Possibilities in the Performance of the CERN Intersecting Storage Rings", Proceedings of the VIth International Conference on High Energy Accelerators, Geneva, Switzerland, August 22, 1967.

42. E. Keil and A.M. Sessler, "Performance Capabilities of Proton Storage Rings", Proceedings of the VIth International Conference on High Energy Accelerators (Cambridge, 1967), p. 77.

43. A.M. Sessler and V.G. Vaccaro, "Passive Compensation of Longitudinal Space Charge Effects in Circular Accelerators", The Helical Insert: CERN Report, Geneva, 1968.

44. E.D. Courant, et al, "Bypass-Storage Ring Option for NAL", Nuclear Instruments and Methods 60, 29 (1968).

45. A.M. Sessler, "Seven Articles in Symposium on Electron Rings Accelerators", Lawrence Radiation Laboratory Report UCRL-18103, 1968: p. 11, p. 137, p. 164, p. 186, p. 191, p. 431, p. 442.

46. D. Keefe, et al., "Experiments on Forming Intense Rings of Electrons Suitable for the Acceleration of Ions", Phys. Rev. Lett. 22, 558 (1969).

47. B.S. Levine and A.M. Sessler, "Excitation of a Closed Cylindrical Cavity by a Charged Ring Moving Along the Axis at Constant Velocity", Proceedings National Particle Accelerator Conference 1969, IEEE Trans. on Nucl. Science NS-16, 3, June 1969, 1031.

48. L.J. Laslett and A.M. Sessler, "A Method for Static-Field Compression in an Electron-Ring Accelerator", Proceedings National Particle Accelerator Conference 1969, IEEE Trans. on Nucl. Science NS-16, Number 3, June 1969, 1034.

49. M.J. Lee, et al., "Beam Amplitude Behavior Upon Crossing a Linear Coupling Resonance With Damping in One Dimension", Proceedings National Accelerator Conference 1969, IEEE Trans. on Nucl. Science ibid, 176.

50. B.V. Chirikov, E. Keil and A.M. Sessler, "Stochasticity in Many-Dimensional Non-Linear Oscillating Systems", Journal of Statistical Physics 3, 307 (1971).

51. C. Pellegrini and A.M. Sessler, "Lower Bounds on Ring Self-Focusing so as to Maintain Ring Integrity During Spillout and Subsequent Acceleration", Nuclear Instruments and Methods 86, 273 (1970).

52. C. Pellegrini and A.M. Sessler, "Crossing of an Incoherent Integral Resonance in the Electron Ring Accelerator", Nuclear Instruments and Methods 84, 109 (1970).

53. R.D. Hazeltine, M.N. Rosenbluth and A.M. Sessler, "Diffraction Radiation by a Line Charge Moving Past a Comb: A Model of Radiation Losses in an Electron Ring Accelerator", Journal of Math. Phys. 12, 502 (1971).

54. C. Pellegrini and A.M. Sessler, "The Equilibrium Length of High-Current Bunches in Electron Storage Rings", IL Nuovo Cimento 3A,116 (1971).

55. E. Keil, C. Pellegrini and A.M. Sessler, "Diffraction Radiation Defocussing of an Electron Ring", Nuclear Instruments and Methods 95, 131 (1971).

56. D. Keefe, et al., "Experiments on Forming, Compressing and Extracting Electron Storage Rings for the Collective Acceleration of Ions", Nuclear Instruments and Methods 93, 541 (1971).

57. A.M. Sessler, "Beam-Surrounding Interactions and the Stability of Relativistic Particle Beams", Proceedings of 1971 Particle Accelerator Conference, IEEE Trans. on Nuclear Science NS-18, No.3, June 1971, p. 1039.

58. A. Faltens, et al., "An Analog for Measuring the Longitudinal Coupling Impedance of a Relativistic Beam With Its Environment", Proc. of the VIIIth International Conference on High Energy Accelerators, 338 (1971).

59. Dieter Mohl and A.M. Sessler, "The Use of RF-Knockout for Determination of the Characteristics of the Transverse Coherent Instability of an Intense Beam", Proceedings of the VIIIth International Conference on High Energy Accelerators, 334 (1971).

60. C. Pellegrini, et al., "A High-Energy Proton-Electron-Positron Colliding Beam System", Proceedings of the VIIIth International Conference on High Energy Accelerators, 153 (1971).

61. J. Druzbick, J.A. White and A.M. Sessler, "Relativistic Contribution to the Hyperfine Interval of the Metastable Triplet State of He 3", Phys. Rev. Lttr. A5, 2683 (1972).

62. D. Mohl, L.J. Laslett and A.M. Sessler, "On the Performance Characteristics of Electron Ring Accelerators", Particle Accelerators 4, 159 (1973).

63. M. Allen, et al., "Status Report on the LBL-SLAC Proton-Electron-Positron Colliding Beam Project", Proceedings III All-Union National Conference on High Energy Particle Accelerators, Moscow, USSR, Oct. 2-4, 1972 , II, p. 292 (1973).

64. D. Mohl, L.J. Laslett and A.M. Sessler, "Transverse Two-Stream Instability in the Presence of Strong Species-Species and Image Forces", Nuclear Instruments 121, 517 (1974).

65. A.M. Sessler, "High-Intensity Effects in the Longitudinal Motion of Stored Particle Beams", in Proceedings of the 1973 Particle Accelerator Conference, IEEE Trans. on Nucl. Sci. NS-20, No. 3, June 1973, p. 854.

66. T. Elioff et al., "Proton-Electron-Positron Design Study", in Proceedings of the 1973 Particle Accelerator Conference, Trans. on Nucl. Sci. NS-20, IEEE No. 3, June 1973, p. 1039.

67. E. Keil, C. Pellegrini and A.M. Sessler, "Tune Shifts for Particle Beams Crossing at Small Angles in the Low-Section of a Storage Ring", Nuclear Instruments and Methods 118, 165 (1974).

68. E. Keil, C. Pellegrini, A. Turrin and A.M. Sessler, "Beam Cavity Interaction in Electron Storage Rings", Nuclear Instruments and Methods 136, 473 (1976).

69. Paul J. Channell and A.M. Sessler, "Strong Turbulence and the Anomalous Length of Stored Particle Beams", Nuclear Instruments and Methods 136, 473 (1976).

70. M.A. Levine, et al., "Tormac Confinement, Theory & Experiment", Plasma Physics Controlled Nuclear Fusion Research 1978, Int. Atomic Energy Agency, Vienna, 1979, p. 81.

71. J. Channell, A. M. Sessler and J. S. Wurtele, "Longitudinal Stability of Intense Non-Relativistic Particle Bunch in Resistive Structures", App. Phy. Lett. 39 (4), 359 (1981).

72. D. Prosnitz and A. M. Sessler, "Millimeter Wave Generator by a Single-Pass, Compton Regime, Variable Parameter Free Electron Laser", Sun Valley 1981, in Free Electron Generators of Coherent Radiation, Addison Wesley, 1982, Reading, Mass., Vol 9, p. 651.

73. C. Litwin, M. Vella and A.M. Sessler, "Linear Electrostatic Stability of the Electron Beam Ion Source", Nuclear Instruments and Methods in Physics Research 198, 189, (1982).

74. A.M. Sessler, "The Free Electron Laser as a Power Source for a High Gradient Accelerating Structure", Proc. of the Workshop on Laser Acceleration of Particles, 1982, American Institute of Physics, Conference Proceedings 91, 154 (1982).

75. J. Channell, J. S. Wurtele and A. M. Sessler, "On the Density Oscillations of a Warm Particle Bunch", Physics of Fluids 26, 2281 (1983).

76. A. Paul, et al, "A Variable Emittance Filter for the Electron Laser Facility", Physics of Fluids 26, 2281 (1983).

77. T. Orzechowski, et al., "The Status of the LBL and LLNL Free Electron Facility", in Free Electron Generators, SPIE 453, 65 (1984).

78. J. Peterson and A.M. Sessler, "Report of the Storage Ring Design Group, Proc. of the Free Electron Generation of Extreme VUV Radiation, 1983", American Institute of Physics, Conference Proceedings 118, p. 266 (1984).

79. D. Hopkins, A.M. Sessler, and J. Wurtele, "The Two-Beam Accelerator", Nuclear Instruments and Methods 228, 15 (1984).

80. T. Orzechowski, et al., "Microwave Radiation from High-Gain Free Electron Laser Amplifier", Phys. Rev. Letters 54, 889 (1985).

81. E.T. Sharlemann, A.M. Sessler, and J. Wurtele, "Optical Guiding in a Free Electron Laser", Proc. of the International Workshop on Coherent and Collective Properties of Relativistic Electron Beams and Electromagnetic Radiation, UCRL-91476, Nuclear Instruments and Methods in Physics Research, A239, 29 (1985).

82. J. Peterson and A.M. Sessler, "Summary of the Working Group on Design of an FEL Storage Ring for l < 1000A o", Proc. of the International Workshop on Coherent and Collective Properties of Relativistic Electron Beams Electromagnetic Radiation, Nuclear Instruments Methods in Physics Research A239, 119 (1985).

83. E.T. Sharlemann, A.M. Sessler, and J. Wurtele, "Optical Guiding in a Free Electron Laser", Phys. Rev. Lett. 54, 1925 (1985).

84. A.M. Sessler, "Report of the Panel Discussion on Future R&D Cooperation", Proc. of the 1984 IFCA Seminar on Future Perspectives in HEP, May 14-20, 1984, KEK Report 84-14, p. 310, Sept. 1984.

85. T. Orzechowski, et al., "High Gain Free Electron Lasers Using Induction Linear Accelerators", IEEE Journal of Quantum Electronics QE21, 831 (1985).

86. A.M. Sessler, "Report of the Working Group on Other Acceleration Schemes", Proceedings of the Second Workshop on the Laser Acceleration of Particles, Malibu, Jan. 1985, American Institute of Physics Conference Proceedings 130, 350 (1985).

87. R. Kuenning, A.M. Sessler, and J.W. Wurtele, "Phase and Amplitude Considerations for the Two-Beam Accelerator", Proceedings of the Second Workshop on the Laser Acceleration Physics Conference Proceedings 130, 324 (1985).

88. W.A. Barletta, et al., "Enhancing the Performance of a High-Gain Free Electron Laser Operating at Millimeter Wavelengths", Nuclear Instruments and Methods in Physics Research A239, 47 (1985).

89. F. Selph and A.M. Sessler, "Transverse Wakefield Effects in the Two-Beam Accelerator", Nuclear Instruments and Methods in Physics Research A244, 323 (1986).

90. E. Sternbach and A.M. Sessler, "A Steady-State FEL: Particle Dynamics in the FEL Portion of a Two-Beam Accelerator", Proceedings of the 7th Intl. FEL Conference, Nuclear Instruments and Methods in Physics Research A250, 464 (1986).

91. J.S. Wurtele, E.T. Scharlemann and A.M. Sessler, "FEL Performance with Two Waveguide Modes", Proc. of the 7th Int. FEL Conf., Nuclear Instruments and Methods in Physics Research A250, 176 (1986).

92. R.W. Kuenning and A.M. Sessler, "Phase and Amplitude Control of the Radio Frequency Wave in the Two-Beam Accelerator", Nuclear Instruments and Methods in Physics Research A243, 263 (1986).

93. T.J. Orzechowski, et al., "High Gain and High Extraction Efficiency from a Free Electron Laser Amplifier Operating in the Millimeter Wave Regime", Proceedings of the 7th International FEL Conference, Nuclear Instruments and Methods in Physics Research A250, 144 (1986).

94. D.B. Hopkins, et al., "High-Power 35 GHz Testing of a Free-Electron Laser and Two-Beam Accelerator Structures", Proc. of the SPIE High Intensity Laser Processes Conference 664, 61, Quebec City, Canada, June 2-6, 1986 (1986).

95. A.M. Sessler and D.B. Hopkins, "The Two-Beam Accelerator", Proceedings of the 1986 Linear Accelerator Conference, June 2-6, 1986, SLAC, 385 (1986).

96. W. M. Fawley, et al., "Novel Accelerator Employed High-Current Electron Beams: Numerical Simulations", Proceedings of the 1986 Linear Accelerator Conference, June 2-6, 1986, SLAC, 112 (1986).

97. J. Masud, et al., "Sideband Control in a Millimeter Wave Free Electron Laser", Phys. Rev. Lttr. 58, 763 (1987).

98. T. Orzechowski, et al., "High Efficiency Extraction of Microwave Radiation From a Tapered Wiggler Free Electron Laser", Phys. Rev. Lttr. 57, 2172 (1986).

99. F.T. Scharlemann, et al.,"Intrinsic Corrections to Optical Guiding in a Free Electron Laser", Ninth International Free-Electron Laser Conference, Williamsburg, Virginia, Sept 14-18, 1987.

100. S.S. Yu, et al., "Waveguide Suppression of the Free Electron Laser Sideband Instability", Proc. of the 8th International Free Electron Laser Conference, Nuclear Instruments & Methods in Physics Research A259, 219 (1987).

101. R.W. Kuenning, A.M. Sessler, and E.J. Sternbach, "Radio Frequency Phase in the FEL Section of a TBA", Proc. of Symposium on Advance Acceleration Concepts, Madison, Aug. 1986 (1987).

102. J.S. Wurtele and A.M. Sessler, "High Repetition Rate Linear Colliders at TeV and GeV Energies", Proc. of Symposium on Advanced Accelerator Concepts, Madison, Aug. 1986 (1987).

103. A.M. Sessler and S.S. Yu, "Relativistic Klystron Version of the Two-Beam Accelerator", Phys. Rev. Lett. 58, 2439 (1987).

104. D.B. Hopkins, et al., "Fabrication and 35 GHz Testing of Key Two-Beam Accelerator Components", Proc. of the IEEE Particle Accelerator Conf., May 16-19, 1987.

105. Y. Goren and A. M. Sessler , "Phase Control of the Microwave Radiation in Free Electron Laser Two-Beam Accelerator", Proc. of the Workshop on New Developments in Particle Acceleration Techniques, Proceedings 1, 231, June 29 - July 4, Orsay France, 1987.

106. A.L. Throop, et al., "Experimental Characteristics of a High-Gain Free- Electron Laser Amplifier Operating at 8-mm and 2-mm Wavelengths", AIAA 19th Fluid Dynamics, Plasma Dynamics and Laser Conference, June 1987.

107. W. A. Barletta and A. M. Sessler, "Radiation from Fine, Intense, Self-Focused Beams at High Energy", Proc. of the I.N.F.N. International School on Electromagnetic Radiation and Particle Beams Acceleration, Varenna, Italy, North-Holland Physics p. 211 (1989).

108. M. A. Allen, et al., "Relativistic Klystron Research and Development" 1988 European Particle Accelerator Conference, Rome, Italy, June 7-11, 1988.

109. D. B. Hopkins, et al., "Design and Fabrication of 33 GHz High-Gradient Accelerator Sections", Presented at the 1988 European Particle Accelerator Conference, Rome, Italy, June 7-11, 1988. LBL-25368.

110. D. B. Hopkins and A. M. Sessler, "Status of LBL/LLNL FEL Research for Two Beam Accelerator Applications", Proceedings of the Workshop on the Physics of Linear Colliders, Capri, Italy, 329 (1989). LBL-26253.

111. M. A. Allen, et al., "Relativistic Klystron Research for Linear Colliders", Presented at DPF Summer Study: Snowmass 1988, High Energy Physics in the 1990's, Snowmass, Colorado, June 27 - July 15, 1988; Proceedings of 1988 Linear Accelerator Conference, Continuous Electron Beam Accelerator Facility, Newport News, VA, October 3-7, 1988 (June 1989).

112. D.H. Whittum, et al., "Plasma Suppression of Beamstrahlung", Proceedings of the Workshop on Physics of Linear Colliders, Capri, Italy, 41, (1989); and also presented at the Conference on Submicron Particle Beam Dynamics and Plasma Acceleration Methods, Santa Barbara, California, July 18-29, 1988, Particle Accelerators $\underline{34}$, 89 (1990), LBL-25759.

113. Yu-Jiuan Chen, E. T. Scharlemann, and A. M. Sessler, "Intrinsic Corrections to Optical Guiding in a Free Electron Laser", Nuclear Instruments and Methods in Physics Research $\underline{A72}$, 485-489 (1988).

114. A.L. Throop, et. al, "Experimental Results of a High Gain Microwave FEL Operating at 140 GHz", Nuclear Instruments and Methods in Physics Research $\underline{A72}$, 15-21 (1988).

115. S. Chattopadhyay, et al., "Conceptual Design of a Bright Electron Injector Based on a Laser-Driven Photocathode RF Electron Gun", Proceedings of 1988 Linear Accelerator Conference, Continuous Electron Beam Accelerator Facility, Newport News, VA, October 3-7, 1988, 325 (June 1989).

116. D.B. Hopkins, et al., "An FEL Power Source for a TeV Linear Collider", Proceedings of 1988 Linear Accelerator Conference, Continuous Electron Beam Accelerator Facility, Newport News, VA, October 3-7, 1988, 684 (June 1989).

117. A.M. Sessler, E. Sternbach and J. S. Wurtele, "A New Version of a Free Electron Laser Two-Beam Accelerator", Presented at the Tenth International Conference on the Application of Accelerators In Research and Industry, Denton, Texas, November 7-9, 1988, Nuclear Instruments and Methods in Physics Research B40/41 (1989) 1064-1068.

118. W.M. Sharp, et al., "Simulation of Superradiant Free-Electron Lasers", Presented at the 10th International Free-Electron Laser Conference, Jerusalem, Israel, August 28- September 2, 1988, Nucl Instr & Methods $\underline{A}$ $\underline{285}$, 217 (1989).

119. D. B. Hopkins, et al., "Elements of a Realistic 17 GHz FEL/TBA Design", Proceedings of the 1989 Workshop on Advanced Accelerator Concepts, Lake Arrowhead, California, AIP Conference Proceedings 193, 141 (1989).

120. D. H. Whittum, et al., "On the Re-Acceleration of Bunched Beams", Proceedings of the 1989 Workshop on Advanced Accelerator Concepts, Lake Arrowhead, California, AIP Conference Proceedings 193, 433 (1989).

121. D. H. Whittum, G. A. Travish, and A. M. Sessler, "Beam Break-Up in the Two Beam Accelerator", Proc. of the 1989 IEEE Particle Accelerator Conference 2, 1190, Chicago, Illinois, March 20-23, 1989.

122. D. B. Hopkins and A. M. Sessler, "Status of LBL/LLNL FEL Research for Two Beam Accelerator Applications", Proc. of the 1989 IEEE Particle Accelerator Conference 2, 1262, Chicago, Illinois, March 20-23, 1989.

123. B. Autin, A. M. Sessler and D. H. Whittum, "Plasma Compensation Effects with Relativistic Electron Beams", Proc. of the 1989 IEEE Particle Accelerator Conference 3, 1812, Chicago, Illinois, March 20-23, 1989.

124. M. A. Allen, et al., "Relativistic Klystrons", Proc. of the Particle Accelerator Conference 2, 1122, Chicago, Illinois, March 20-23, 1989.

125. J. Dawson and A. Sessler, "DNA Base Pair Sequencer: Scanning Tunneling Microscope Plus Infrared Radiation", Proceedings of Workshop on X-Ray Microimaging for the Life Sciences, LBL-27660, UC-600, CONF-8905192, 108, August 1989.

126. M. A. Allen, et al., "High-Gradient Electron Accelerator Powered by a Relativistic Klystron", Phys. Rev. Lett. 63, 2472 (1989).

127. P. Chen, K. Oide, A. M. Sessler and S. S. Yu, "An Adiabatic Focuser", Proceedings of the XIV International Conference on High Energy Accelerators, Tsukuba, Particle Accelerators 31, 7 (1990).

128. A. M. Sessler, D. H. Whittum and J. S. Wurtele, "A Study of Phase Control in the FEL Two-Beam Accelerator", Proceedings of the XIV International Conference on High Energy Accelerators, Tsukuba, Particle Accelerators 31, 69 (1990).

129. R. A. Jong, et al., "17.1-GHz Free-Electron Laser as a Microwave Source for TeV Colliders", Presented at the International FEL Conference, Florida, August, 1989, Nucl Instr and Meth A296, 776 (1990).

130. M.A. Allen, et al., "Recent Progress in Relativistic Klystron Research", Proceedings of the XIV International Conference on High Energy Accelerators, Tsukuba, Particle Accelerators 30, 189 (1990).

131. P.Chen, K. Oide, A.M. Sessler and S.S. Yu, "A Plasma-Based Adiabatic Focuser", Phys. Rev. Letts. 64, 1231 (1990).

132. C. Joshi, C.E. Clayton, K. Marsh, D.B. Hopkins, A. Sessler and D. Whittum, "Demonstration of the Frequency Upshifting of Microwave Radiation by Rapid Plasma Creation", IEEE Trans. on Plasma Science 18 No. 5, 814, Oct. 1990.

133. D. Whittum, A.M. Sessler and J.M. Dawson, "The Ion-Channel Laser", Phys. Rev. Letts. 64, 2511 (1990).

134. D. H. Whittum, A.M. Sessler and V.K. Neil, "Transverse Resistive Wall Instability in the Two-Beam Accelerator", Phys. Rev. A, 43, No. 1, 294 (1991).

135. M.A. Allen, et al, "RF Power Sources for Linear Colliders", SLAC-PUB 5274, LBL-29458, June 1990, contributed to 2nd European Particle Accelerator Conference, Nice, France, June 12-16, 1990.

136. G.A. Westenskow, et al, "Relativistic Klystrons for High-Gradient Accelerators", Proceedings of the 1990 Linear Accelerator Conference, Albuquerque, LA-12004-C Conference, 192 (1991).

137. W.M. Sharp, et al, "Simulation of a Standing-Wave Free-Electron Laser", Proceedings of the 1990 Linear Accelerator Conference, Albuquerque, LA-12004-C Conference, 656 (1991), UCRL-JC-103826.

138. D.I. Sessler, J. McGuire and A.M. Sessler, "Perioperative Thermal Insulation", Anesthesiology 74, 875 (1991).

139. W.M. Sharp, G. Rangarajan, A.M. Sessler and J.S. Wurtele, "Phase Stability of a Standing-Wave Free-Electron Laser", Proceedings of Intl. Soc. Opt. Eng. (SPIE), Jan. 20-25, 1991, UCRL-JC-105569.

140. A.M. Sessler, D.H. Whittum, J.S. Wurtele, W.M. Sharp and M.A. Makowski, "Standing-Wave Free-Electron Laser Two-Beam Accelerator", Nucl. Instr. & Meth. in Phys. Res. A 306, 592 (1991).

141. A.M. Sessler, D.H. Whittum and L-H. Yu, "Radio-Frequency Beam Conditioner for Fast-Wave Free-Electron Generators of Coherent Radiation", Phys. Rev. Lett. 68, 309 (1992).

142. K-J. Kim and A.M. Sessler, "Photon Storage Cavities", 13th Free-Electron Laser Conference, Santa Fe, NM, Aug. 25-30, 1991, LBL-31195, Nucl. Instr. & Meth. in Phys. Res. A318, 895 (1992).

143. S. Krishnagopal, M. Xie, K-J. Kim and A. Sessler, "Three-Dimensional Simulation of a Hole-Coupled FEL Oscillator", 13th Free-Electron Laser Conference, Santa Fe, NM, Aug. 25-30, 1991, LBL-31196, Nucl. Instr. & Meth. in Phys. Res. A318, 661 (1992).

144. G. Rangarajan, A. Sessler and W.M. Sharp, "Discrete Cavity Model of a Standing-Wave Free-Electron Laser", 13th Free-Electron Laser Conference, Santa Fe, NM, Aug. 25-30, 1991, LBL-31197, Nucl. Instr. & Meth. in Phys. Res. A318, 745 (1992).

145. L-H. Yu, A. Sessler and D.H. Whittum, "Free-Electron Laser Generation of VUV and X-Ray Radiation Using a Conditioned Beam and Ion-Channel Focusing", 13th Free-Electron Laser Conference, Santa Fe, NM, Aug. 25-30, 1991, LBL-31198, Nucl. Instr. & Methods in Phys. Res. A 318, 721 (1992).

146. K. Takayama, R. Govil and A. Sessler, "Macroparticle Theory of a Standing Wave Free-Electron Laser Two-Beam Accelerator", Nucl. Instr. & Meth. in Phys. Res. A320, 587 (1992), LBL-32044.

147. G. Rangarajan and A. Sessler, "Sensitivity Studies of a Standing-Wave Free-Electron Laser", Third Workshop on Advanced Accelerator Concepts, Pt. Jefferson, NY, June 1992, LBL-32463, AIP Conference Proceedings 279, 156, American Institute of Physics, New York (1993).

148. J.S. Wurtele, D.H. Whittum and A.M. Sessler, "Common Analysis of the Relativistic Klystron and the Standing-Wave Free-Electron Laser Two-Beam Accelerator", XVth Intl. Conference on High Energy Accelerators, Hamburg, Germany, July 20-24, 1992, Int. J. Mod. Phys. A, 2A, 508, (1993), LBL 32580.

149. D.I. Sessler, A.M. Sessler, S. Hudson, A. Moayeri, "Heat Loss During Surgical Skin Preparation", submitted to Anesthesiology 7/30/92.

150. S. Krishnagopal, G. Rangarajan and A. Sessler, "The Multi-Cavity Free-Electron Laser", LBL-32220, Optics Communications 100, 518 (1993).

151. W. Barletta, A. Sessler and L.-H. Yu, "Physically Transparent Formulation of a Free-Electron Laser in the Linear Regime", 14th Intl. Free Electron Laser Conference, Kobe, Japan, Aug. 23-28, 1992, LBL-32221, Nucl. Instr. & Meth. in Phys. Res. A, 331, 491 (1993).

152. R. Govil, R.A. Rimmer, A. Sessler and H.G. Kirk, "Design of RF Conditioner Cavities", 14th Intl. Free Electron Laser Conference, Kobe, Japan, Aug. 23-28, 1992, LBL-32232, Nucl. Instr. & Meth. in Phys. Res. A, 331, 335 (1993).

153. I. Ben-Zvi, L-H. Yu, R. Govil and A.M. Sessler, "A Proposed Experiment for Beam Conditioning", 14th Intl. Free Electron Laser Conference, Kobe, Japan, Aug. 23-28, 1992, BNL-47857, Nucl. Instr. & Meth. in Phys. Res. A, 331, ABS 1 (1993).

154. D.K. Kalluri, V.R. Gotenti and A. Sessler, "WKB Solution for Wave Propagation in a Time-Varying Magnetoplasma Medium: Longitudinal Propagation", IEEE Trans. on Plasma Science, Vol. 21, No. 1, p. 70 (Feb. 1993).

155. J. Wurtele, D. Whittum and A. Sessler, "Impedance-Based Analysis and Study of Phase Sensitivity in Slow-Wave Two-Beam Accelerators", Third Workshop on Advanced Accelerator Concepts, Pt. Jefferson, NY, June 1992, LBL 31848, AIP Conference Proceedings 279, 143, American Institute of Physics, New York (1993).

156. K. Belani, D.I. Sessler, A.M. Sessler, et al, "Leg Heat Content During the Core Temperature Plateau", Anesthesiology 78, 856 (1993).

157. S. Krishnagopal and A. M. Sessler, "Stability of Resonator Configurations in the Presence of Free-Electron Laser Interactions", to be published in Optics Communications, LBL-33053.

158. W. Barletta and A. Sessler, "Characteristics of a High Energy $\mu^+\mu^-$ Collider Based on Electro-production of Muons", published Nuc. Instr. & Meth. in Phys. A, 350 (1994) 36, LBL-33613, UCRL-JC-113093.

159. S. Krishnagopal, G. Rangarajan and A. Sessler, "Numerical Studies of the Multi-Cavity Free-Electron Laser", Proceedings of the 14th International Free Electron Laser Conference, Kobe, Japan, Aug. 23-28, 1992, Nucl. Instr. & Meth. in Phys. Res. A, 331, ABS 22 (1993).

160. R. Govil and A. Sessler, "Theoretical Examination of Transfer Cavities in a Standing-wave Free-electron Laser Two-beam Accelerator", *Intense Microwave Pulses*, Howard E. Brandt, Ed., Proc. SPIE 1872, 130 (1993), LBL-32480.

161. C. Wang and A. Sessler, "Three-Dimensional Simulation Analysis of the Standing-wave Free-electron Laser Two Beam Accelerator", *Intense Microwave Pulses*, Howard E. Brandt, Ed., Proc. SPIE 1872, 135 (1993), LBL-32481.

162. J.S. Kim, H. Henke, A.M. Sessler and D.H. Whittum, "Multi-Bunch Beam-Break-Up Studies for a SWFEL/TBA", Proceedings of the 1993 Particle Accelerator Conference, Washington, DC, 3288(1993), LBL-33255.

163. Changbiao Wang and Andrew M. Sessler, "Three-Dimensional Simulation Analysis of the First Sections of a Standing-Wave Free-Electron Laser Two-Beam Accelerator", Proceedings of the 1993 Particle Accelerator Conference, Washington, DC, 2608 (1993), LBL-33256.

164. J-S. Kim, H. Henke, A.M. Sessler and W.M. Sharp, "The Standing Wave FEL/TBA: Realistic Cavity Geometry and Energy Extraction", Proceedings of the 1993 Particle Accelerator Conference, Washington, DC, 2593 (1993), LBL-33257.

165. J. Wei, X-P. Li and A.M. Sessler, "Crystalline Beam Ground State", Proceedings of the 1993 Particle Accelerator Conference, Washington, DC, 3527 (1993).

166. W.M. Fawley, A.M. Sessler and E.T. Scharlemann, "Coherence and Linewidth Studies of a 4-nm High Power FEL", Proceedings of the 1993 Particle Accelerator Conference, Washington, DC, 1530 (1993).

167. W. Barletta, et al, "Plasma Lens Experiments at the Final Focus Test Beam", Proceedings of the 1993 Particle Accelerator Conference, Washington, DC, 2638 (1993).

168. C. Wang and A.M. Sessler, "An Efficient Microwave Power Source: Free-Electron Laser Afterburner", submitted to J. Appl. Phys., LBL-33755.

169. A.M. Sessler and D.H. Whittum, "Suppression of Beamstrahlung by Means of a Plasma", Third Workshop on Advanced Accelerator Concepts, Pt. Jefferson, NY, June 1992, AIP Conference Proceedings 279, 939, American Institute of Physics, New York (1993).

170. S. Krishnagopal and A.M. Sessler, "Generation of Harmonic Radiation Using the Multi-Cavity Free-Electron Laser", Proceedings of the 15th Intl. Free Electron Laser Conference, The Hague, The Netherlands, August 1993, 331 (1994) LBL 34015.

171. R.L. Mills and A.M. Sessler, "Liouville's Theorem and Phase-Space Cooling", Proceeding of the Workshop on Beam Cooling and Related Topics, Montreux, Switzerland, October 3-8, 1993, CERN Report 94-03, 4 (1994).

172. W.A. Barletta and A.M. Sessler, "Stochastic Cooling in Muon Colliders", Proceeding of the Workshop on Beam Cooling and Related Topics, Montreux, Switzerland, October 3-8, 1993, LBL 34680, p. 145.

173. J. Wei, X-P. Li and A.M. Sessler, "Critical Temperatures for Crystalline Beam", Workshop on Beam Cooling and Related Topics, Montreux, Switzerland, October 3-8, 1993.

174. W. Barletta, S.S. Gershtein, V. Krishan, M. Reiser, A.M. Sessler, M. Xie, "Collective Acceleration in Solar Flares", published in Proceedings of Kardamili Conference, Kardamili, Greece, Aug. 29-Sept. 4, 1993, Physica Scripta, Vol. T52 (1994), p.154.

175. J. Wei, X-P. Li and A.M. Sessler, "The Low Energy States of Circulating Stored Ion Beams: Crystalline Beams", published in Physical Review Letters, Dec. 5, (1994), vol. 73, No. 23, 3089, LBL-35322.

176. H. Okamoto, A.M. Sessler and D. Mohl, "Three Dimensional Laser Cooling of Stored and Circulating Ion Beams by Means of a Coupling Cavity", to be published in Physical Review Letters, 72, 3977 (1994).

177. A. Sessler, "Theory of the Standing-Wave Free-Electron Laser Two-Beam Accelerator", Proceedings of the International Workshop on Pulsed RF Power Sources for Linear Colliders, Dubna-Protvino, Russia, 279 (1993).

178. Xiao-Ping, Li, Jie Wei, A.M. Sessler, "Crystalling Beams in a Storage Ring: How Long Can It Last?", published in the Proceeding of the EPAC'94, London, UK, June 27-July 1, 1994.

179. D. Mohl, A.M. Sessler, "Report of the Working Session on Cyclotron Maser Cooling", Proceeding of the Workshop on Beam Cooling and Related Topics", Montreux, Switzerland, Oct. 4-8, 1994, p. 429.

180. S. Chattopadhyay, W. Barletta, S. Maury, D. Neuffer. A. Ruggiero, A. Sessler, "Critical Issues in Muon Colliders - A Summary", Proceeding of the Workshop on Beam Cooling and Related Topics, Montreux, Switzerland, Oct. 4-8, 1994, p. 439.

181. T. Matsukawa, M.D., D. Sessler, M.D., A. Sesssler, Ph.D., M. Schroeder, B.A., M. Ozaki,M.D., A. Kurz, M.D., C. Cheng, M.D., "Heat Flow and Distribution During Induction of General Anesthesia", submitted Anesthesiology.

182 J. Wei, X-P. Li, A.M. Sessler, "Crystalline Beam Ground State", RHIC Project, Brookhaven National Laboratory, BNL #52381, June 11, 1993.

183 M. Xie, K-J. Kim, A. M. Sessler, "Crab Crossing in a Gamma-Gamma Collider", Published in the Proceeding of the Gamma-Gamma Collider Workshop, March 1994.

184. H.-D. Nuhn, A. M. Sessler, et. al., "Short Wavelength FELs Using the SLAC Linac", published in the Proceeding of the EPAC'94, London, England, June 27-July 1, 1994.

185. S. Yu, F. Deadrick, N. Goffeney. E. Henestroza, T. Houck, H. Li, C. Peters, L. Reginato, A. Sessler, D. Vanecek, G. Westenskow, "Relativistic-Klystron Two-Beam-Accelerator As A Power Source For a 1 TeV Next Linear Collider - A System Study", presented at LINAC'94, August 20-26, 1994, Tsukuba, Japan.

186. H. Li, S. Yu and A. Sessler, "Design Study of Longitudinal Dynamics of the Drive Beam in a Relativistic Klystron Two-Beam Accelerator", presented at and to be published in the Thirty-Sixth Annual Meeting of the American Physical Society, Minneapolis, MN, November 7-11, 1994, LBL # 36180.

187. H. Li, T. Houck, N. Goffeney, E. Henestroza, A. Sessler, G. Westenskow and S. Yu. "Design Study of Beam Dynamics Issues for 1 TeV Next Linear Collider Based Upon the Relativisitic Klystron Two-Beam Accelerator", presented at and to be published in the proceeding of the Advanced Accelerator Concepts Conference, Lake Geneva, WI, June 12-18, 1994. LBL #36233.

188. W. Barletta, S. Chattopadhyay, P. Chen, D.Cline, A. Sessler et. al, "Plasma Lens Experiments at the Final Focus Test Beam", presented at the Advanced Accelerator Concepts Conference, June 12-18, 1994, Lake Geneva, WI.

189. B. Ivanov, V. Butenko, A. Egorov, and A. Sessler et.al. "Development of a Project and Theoretical Substantiations of a Two-Beam Electron-Ion Accelerator Based on Doppler Effect", presented at the Advanced Accelerator Concepts Conference, June 12-18, 1994, Lake Geneva, WI.

190. H. Okamoto, A. Sessler, and D. Mohl," Three-Dimensional Laser Cooling", published in the Proceeding of the EPAC'94, London, England, June 27-July 1, 1994.

191. Jie Wei, Xiao-Ping Li, Andrew M. Sessler, "Crystalline Beam Properties As Predicted For the Storage Rings ASTRID and TSR", presented at the 1995 PAC, Dallas, TX, May 1-5, 1995.

192. P. Chen, A. Sessler, et.al., "Progress On Plasma Lens Experiments at the Final Focus Test Beam", presented at the 1995 PAC, Dallas, TX, May 1-2, 1995.

193. G. Giordano, H. Li, N. Goffeney, E. Henestroza, A. Sessler, S. Yu, "Beam Dynamics Issues in an Extended Relativistic Klystron", presented at the 1995 PAC, Dallas, TX, May 1-5, 1995.

194. D. Robin, C. Kim, A. Sessler, "Compton Scattering in the ALS Booster", presented at the 1995 PAC, Dallas, TX, May 1-5, 1995.

Review Papers

1. A.M. Sessler, "Theory of Liquid 3 He", Liquid Helium, Proceedings of the International School of Physics " Enrico Fermi" Course 21 (Academic Press), p. 188 (1963).

2. A.M. Sessler, "Helium Three", Proceedings of the Eighth International Conference on Low Temperature Physics, (edited by Davies, Butterworths), p. 11 (1963).

3. A.M. Sessler, "Instabilities of Relativistic Particle Beams", Proceedings of the Vth International Conference on High Energy Accelerators (Frascati and Roma, 1966), p. 319.

4. A.M. Sessler, "Summary Paper on Beam Behavior", Proceedings of the International Symposium on Electron and Positron Rings, Saclay, Sept.1966 (Presses Universitaires de Frances, 1966), IX.I.

5. A.M. Sessler and Allison et al., "The Electron-Ring Accelerator Program LRL", Proc. of Soviet National Conference on Accelerators 1, 54, Moscow (1970).

6. A.M. Sessler, "The Electron Ring Accelerator", in Encyclopedia of Science and Technology 1969 (McGraw-Hill), p. 252.

7. A.M. Sessler, "The Acceleration of Particles by Collective Fields I", Comments on Nuclear and Particle Physics 3, 93 (1969).

8. A.M. Sessler, "Collective-Field Acceleration", Proceedings VIIth. International Particle Accelerator Conference, Yerevan, 1969, 2, 431 (1970).

9. C. Pellegrini and A.M. Sessler, "Physics With and Physics of Colliding Electron Beams", Comments on Nuclear and Particle Physics, 4, 55 (1970).

10. J.M. Peterson, et al., "The Electron-Ring Accelerator Program at Berkeley, Lawrence Radiation Laboratory Report", UCRL-20150, Proc. of 2nd National Conference on Particle Accelerators, USSR, (Nov. 1970).

11. L.J. Laslett and A.M. Sessler, "The Acceleration of Particles by Collective Fields II", Comments on Nuclear and Particle Physics 4, 211 (1970).

12. A.M. Sessler, "The Possibilities and Problems of Present and Future Accelerators", Proceedings of the VIIth International School on Particle Physics, Yerevan (Nov. 1971).

13. A.M. Sessler, "Instabilities of Relativistic Particle Beams", Proceedings of the VIIth International School on Particle Physics, Yervan, (Nov. 1970).

14. A.M. Sessler, "Collective Phenomena in Accelerators", in Proceedings of the Proton Linear Accelerator Conference, Los Alamos Scientific Laboratory LA-5115, p. 291 (1972).

15. A.M. Sessler, "Wigglers and Free Electron Lasers", An Overview, Wiggler Magnets, (Edited by H. Winick and T. Knight), SSRP Report 77/05 (May 1977).

16. D. Keefe and A.M. Sessler, "Heavy Ion Fusion", Proc. XI International Conference on High Energy Accelerators, CERN Geneva, 1981, p. 201.

17. A.M. Sessler, "Collective Field Accelerators in Physics of High Energy Particle Accelerators", AIP Conference Proceedings No. 87, p. 919 (1982).

18. A.M. Sessler, "Report of the Working Group on Media Accelerators", Proc. of the Workshop on Laser Acceleration of Particles, 1982, AIP, Conference Proceedings No.91, p. 10 (1982).

19. A.M. Sessler, "The Research Needs of the New Accelerators Technologies", Proc. of the Seminar on New Trends in Particle Acceleration Techniques, June 1-3, 1982, LBL-14819.

20. A.M. Sessler, "Collective Effect Accelerators and the Challenge of Attaining Ultra-High Energies", Proc. of the ECFA-RAL Meeting on the Challenge of Ultra-High Energies, New College, Oxford, Sept. 1982, p. 79.

21. A.M. Sessler, "Laser Accelerator", IEEE Trans. on Nucl. Sci. NS-30, No.4, 3145 (1983).

22. A.M. Sessler, "New Concepts in Particle Accelerators", Proc. of the XIIth International Conference on High Energy Accelerators, Fermilab, 1983, p. 445.

23. A.M. Sessler, "Topics in the physics of Particle Accelerators", in Techniques and Concepts of High-Energy Physics III, (edited by T. Ferbel, Plenum Press), New York (1985), p. 337.

24. A.M. Sessler, "New Acceleration Methods", in Techniques and Concepts of High-Energy Physics III, (edited by T. Farbel, Plenum Press), New York (1985), p. 375.

25. A.M. Sessler, "The Quest For Ultra-High Energies", American Journal of Physics 54, 505 (1986).

26. W. Colson and A.M. Sessler, "The Free Electron Laser", Annual Reviews of Nuclear and Particle Science 35, 25 (1985).

27. A.M. Sessler, "Future Accelerator Technology", Proceeding of the 2nd Conference on the Intersections between Particle and Nuclear Physics, May 26-31, 1986, Lake Louise, American Institute of Physics Conference Proceedings 150, 53 (1986).

28. A.M. Sessler and D. Vaughan, "Free Electron Laser", American Scientist 75, 34 (1987).

29. A.M. Sessler and W. Schnell, "Semi-Conventional High-Frequency Linacs", Summary Report of Working Group I, Workshop on New Development in Particle Acceleration Techniques, Proceedings 1, 137, June 29 - July 4, Orsay France, 1987.

30. A.M. Sessler, "New Particle Acceleration Techniques", Physics Today. 41, No. 1, January 1988, p 26-34.

31. A.M. Sessler, "The Two-Beam Accelerator", Proceedings of the International Conference High Energy Physics, 1987, Uppsala.

32. N. Bloembergen, et al., "Report to the APS by the study group on Science and Technology of Directed Energy Weapons", Rev. of Mod. Phys. 59, No 3 Part II (1987).

33. A.M. Sessler, "Report of the New Acceleration Schemes Panel", Proc. of the ICFA Meeting on New Perspectives in HEP, Brookhaven, Oct. 1987.

34. A. M. Sessler, "The Two-Beam Accelerator and the Relativistic Klystron Power Source" International Memorial Seminar for G. I. Budker, Novosibirsk, U.S.S.R., April 26-30, 1988.

35. A. M. Sessler, "Pulsed Power Technology and High Energy Particle Accelerators" Whats New in Physics, Physics Today 42, S-39 January, 1989.

36. A. M. Sessler, "High-Power, High-Efficiency FELs", Proceedings of the CERN Accelerator School on Synchrotron Radiation and Free Electron Lasers, Chester College, Chester, England, April 6-13, 1989, CERN 90-03, p. 335 (1990).

37. A. M. Sessler, "Prospects for the FEL", Proceedings of the CERN Accelerator School, Chester College, Chester, England, April 6-13, 1989, CERN 90-03, p. 373 (1990).

38. A.M. Sessler, "Some Frontiers of Accelerator Physics", Proceedings of the Kent M. Terwilliger Memorial Symposium, University of Michigan, October 13-14, 1989, AIP Conference Proceedings 237, 92, American Institute of Physics, NY (1991).

39. A.M. Sessler, "Frontiers of Particle Beam Physics", 31st Annual Meeting, American Physical Society-Division of Plasma Physics, Anaheim, November 1989, Phys. Fluids B—Plasma Physics 2, No. 6, p. 1325, June (1990).

40. A.M. Sessler, "Beam Dynamics Issues of High-Luminosity Asymmetric Collider Rings, Particle World 1, No. 5, p. 125, 1990.

41. K.-J. Kim and A. Sessler, "Free-Electron Lasers: Present Status and Future Prospects", SCIENCE 250, p. 88, (5 October 1990).

42. A.M. Sessler, "New Techniques for Particle Accelerators", LBL-29114A, presented at 2nd European Accelerator Conference, Nice, France, June 12-16, 1990; Recent Projects and Developments in Acceleration Machines, Varenna, 289, June 1990.

43. A.M. Sessler, "Some Nonlinear Problems in the Manipulation of Beams", LBL-29716, US-Japan Workshop on Nonlinear Dynamics and Acceleration Mechanisms, Tsukuba, Japan, October 22-25, 1990, Nonlinear Dynamics and Particle Acceleration, AIP Conference Proceedings 230, 165, American Institute of Physics, NY (1991).

44. A.M. Sessler, "Beam Conditioning for Free-Electron Lasers", Workshop on Fourth Generation Light Sources, Stanford Synchrotron Radiation Laboratory, Feb. 24-27, 1992, SSRL 92/02, 278 (1992).

45. W.A. Barletta, A.M. Sessler and L.-H. Yu, "Using the SLAC Two Mile Accelerator for Powering an FEL", Workshop on Fourth Generation Light Sources, Stanford Synchrotron Radiation Laboratory, Feb. 24-27, 1992, SSRL 92/02, 376 (1992).

46. A. Sessler, "The Development of Colliders", Intl. Symp. on 30 Years of Neutral Currents: from Weak Neutral Currents to the (W)/Z and Beyond, Santa Monica, CA, Feb. 3-5, 1993, LBL-33664.

47. A.M. Sessler, "Progress in Advanced Accelerator Concepts", presented at the LINAC'94, Tsukuba, Japan, August 21-16, 1994.

48. J. Wei, X-P. Li, A. Sessler, "Crystalline Beams", presented at the Advanced Accelerator Concepts Conference, Lake Geneva, WI, June 12-18, 1994.

49. S. Chattopadhyay, W. Barletta, S. Maury, D. Neuffer, A. Ruggiero, A. Sessler, "Critical issues in low energy muon colliders — a summary", Nuc. Instr. & Meth. in Phys. A, 350 (1994) 53-56.

50. A. M. Sessler, "The Cooling of Particle Beams", to be presented at the Tamura Symposium on Accelerator Physics, Univ. of Texas Austin, Nov. 14-16, 1994.

51. S. Chattopadhyay and A. Sessler, "Linear Colliders with gamma—gamma collisions — an introduction", published in Nucl. Instr. and Meth., February 1, 1995, Vol. 355, No. 1.

52. A.M. Sessler, "Photon-Photon Colliders", presented at the 1995 PAC, Dallas, TX, May 1-5, 1995.

Book Reviews

1. A.M. Sessler, Book Review of Linear Accelerators (by P. Lapostolle, S. Septier), Nuclear Instruments and Methods 95, No. 395 (1971).

2. A.M. Sessler Book Review of Collective Ion Acceleration, (by Olson, Schumacher), Collective Methods of Acceleration, Edited by Rostoker and Reiser, Nuclear Instruments and Methods 171, 627 (1980).

3. A.M. Sessler Book Review of Refusenik (by Mark Ya Azbel), in Bulletin of Atomic Scientists, Vol. 39, No. 7, p. 4 (1983).

4. A.M. Sessler Book Review of Artificial Particle Beams in Space Plasma Studies for American Scientists 73, No. 76 (1985).

5. A.M. Sessler Book Review of Times' Arrow, (by Richard Morris), American Scientists 73, No. 4, 384 (1985).

6. A.M. Sessler Book Review of Relativity and Engineering, (by J. Van Bladel), Physics Today 38, No. 78 (June 1985).

7. A.M. Sessler Book Review of The Periodic Table, (by P. Levi), S.F. Chronicle, April 17, 1985.

8. A.M. Sessler Book Review of Dangerous Thoughts: Memoirs of a Russian Life (by Y. Orlov), for Physics Today.

Books

1. Beam Dynamics Issues of High-Luminosity Asymmentric Collider Rings; AIP 214 Conference Proceedings, Berkeley, CA, 1990, Editor: Andrew M. Sessler

2. The Development of Colliders; Claudio Pellegrini and Andrew M. Sessler, AIP Reprint to be published 1994.

3. Proceeding of the Gamma-Gamma Collider Workshop; Swapan Chattopadhyay and Andrew Sessler, Nuclear Instruments and Methods Section A, published February 1, 1995, Vol. 355, No.1. Workshop held at LBL, March 28-31, 1994.

4. Proceeding of the Advanced Accelerator Concepts Workshop; AIP, Paul Schoessow and Andrew Sessler, to be published 1994.

5. Proceeding of the Workshop on Applications of Accelerators, W.B. Herrmannsfeldt, Andrew M. Sessler, and J.R. Alonzo, December 1-2, 1993, revised January 21, 1994, SLAC-430/LBL-35023.

Other

1. A.M. Sessler, "A View From Home", Nature 263, 624 (1976).

2. A.M. Sessler, The Lawrence Berkeley Laboratory: "A Welcome to the Particles and Radiation Therapy (Part II)", International Conference - Int. J. Radiation Oncology, Biol. Phys. 3, 1 (1977).

3. Rita LaBrie and A.M. Sessler, "Impact on World Science of Soviet Science as Measured by Journal Citations", LBL-16158 (1983).

4. A.M. Sessler and Yvonne Howell, "Andrei Sakahrov: A Man of Our Times", American Journal of Physics 52, 397 (1984).

5. A.M. Sessler, "Reflections occasioned by the release of Yuri Orlov", Editorial, Physics Today 39, 168 (Nov. 1986).

6. A.M. Sessler, "The ELF Facility", LLNL Brochure (1987).

7. A.M. Sessler, "Testimony on the SSC before the U.S. Senate Committee on energy and Natural Resources and Development; Subcommittee on Energy Research and Development", on April 12, 1988.

8. A.M. Sessler, Opening Remarks, 1987 Fermilab Industrial Affiliates Roundtable On Research Technology in the Twenty-First Century, Batavia, Illinois, May 21-22, 1987, 131-135.

9. A. M. Sessler, Remarks, Upon receipt of the Achievement in Accelerator Physics and Technology Prize, July 19, 1988.

10. A.M. Sessler, The physics of beams: Past, present, future, Opinion, Physics Today 43, 69 (June 1990).

11. A.M. Sessler, Introductory Remarks at Yelena Bonner Lecture, Sakharov Symposium, University of California, Berkeley, March 22, 1990; to be published in Proceedings of 1st Intl. A.D. Sakharov Conference on Physics, Moscow, USSR, May 27-31, 1991.

12. A.M. Sessler, ed., Proceedings of Workshop on Beam Dynamics Issues of High-Luminosity Asymmetric Collider Rings, Berkeley, CA , February 12-16, 1990, AIP Conference Proceedings 214, American Institute of Physics, New York (1990).

13. U.S. Patent No. 4,975,655, Dec. 4, 1990, Method and Apparatus for Upshifting Light Frequency by Rapid Plasma Creation, J.M. Dawson, C. Joshi, W.B. Mori, A.M. Sessler and S.C. Wilks.

14. A.M. Sessler, "Accelerator Beam Emittance", Am. J. of Phys., 60, 760 (1992).

15. M. Furman, R. Warnock, F. Izrailev, M. Lieberman and A. Sessler, "Jeffrey Tennyson Remembered", Stanford Linear Accelerator Center Beam Line Vol. 22, No. 3, p. 48 (Fall 1992).

16. A.M. Sessler and M. Pripstein, "S.O.S. and the Eternal Struggle for Human Rights", to be published in the Halback Symposium Proceeding, Lawrence Berkeley Laboratory, 2/3-5/95.

17. S. Yu and A.M. Sessler "Low-Field Permanent Magnet Quadrupoles in a New Relativistic-Klystron Two-Beam Accelerator Design", to be published in the Halbach Symposium Proceeding, Lawrence Berkeley Laboratory, 2/3-5/95.

Author Index

B

Barletta, W. A., 56

C

Chen, P., 68

D

Dawson, J. M., 17

H

Hyde, E., 96

P

Pellegrini, C., 49
Pripstein, M., 120

S

Sessler, A. M., 125
Shvets, G., 24
Siemann, R. H., 1
Symon, K. R., 79

W

Whittum, D. H., 101
Wurtele, J. S., 24

AIP Conference Proceedings

		L.C. Number	ISBN
No. 324	Twelfth Symposium on Space Nuclear Power and Propulsion (Albuquerque, NM 1995)	94-73603	1-56396-427-9
No. 325	Conference on NASA Centers for Commercial Development of Space (Albuquerque, NM 1995)	94-73604	1-56396-431-7
No. 326	Accelerator Physics at the Superconducting Super Collider (Dallas, TX 1992-1993)	94-73609	1-56396-354-X
No. 327	Nuclei in the Cosmos III Third International Symposium on Nuclear Astrophysics (Assergi, Italy 1994)	95-75492	1-56396-436-8
No. 328	Spectral Line Shapes, Volume 8 12th ICSLS (Toronto, Canada 1994)	94-74309	1-56396-326-4
No. 329	Resonance Ionization Spectroscopy 1994 Seventh International Symposium (Bernkastel-Kues, Germany 1994)	95-75077	1-56396-437-6
No. 330	E.C.C.C. 1 Computational Chemistry F.E.C.S. Conference (Nancy, France 1994)	95-75843	1-56396-457-0
No. 331	Non-Neutral Plasma Physics II (Berkeley, CA 1994)	95-79630	1-56396-441-4
No. 332	X-Ray Lasers 1994 Fourth International Colloquium (Williamsburg, VA 1994)	95-76067	1-56396-375-2
No. 333	Beam Instrumentation Workshop (Vancouver, B. C., Canada 1994)	95-79635	1-56396-352-3
No. 334	Few-Body Problems in Physics (Williamsburg, VA 1994)	95-76481	1-56396-325-6
No. 335	Advanced Accelerator Concepts (Fontana, WI 1994)	95-78225	1-56396-476-7 (Set) 1-56396-474-0 (Book) 1-56396-475-9 (CD-Rom)
No. 336	Dark Matter (College Park, MD 1994)	95-76538	1-56396-438-4

No. 337	Pulsed RF Sources for Linear Colliders (Montauk, NY 1994)	95-76814	1-56396-408-2
No. 338	Intersections Between Particle and Nuclear Physics 5th Conference (St. Petersburg, FL 1994)	95-77076	1-56396-335-3
No. 339	Polarization Phenomena in Nuclear Physics Eighth International Symposium (Bloomington, IN 1994)	95-77216	1-56396-482-1
No. 340	Strangeness in Hadronic Matter (Tucson, AZ 1995)	95-77477	1-56396-489-9
No. 341	Volatiles in the Earth and Solar System (Pasadena, CA 1994)	95-77911	1-56396-409-0
No. 342	CAM-94 Physics Meeting (Cacun, Mexico 1994)	95-77851	1-56396-491-0
No. 343	High Energy Spin Physics Eleventh International Symposium (Bloomington, IN 1994)	95-78431	1-56396-374-4
No. 344	Nonlinear Dynamics in Particle Accelerators: Theory and Experiments (Arcidosso, Italy 1994)	95-78135	1-56396-446-5
No. 345	International Conference on Plasma Physics ICPP 1994 (Foz do Iguaçu, Brazil 1994)	95-78438	1-56396-496-1
No. 346	International Conference on Accelerator-Driven Transmutation Technologies and Applications (Las Vegas, NV 1994)	95-78691	1-56396-505-4
No. 347	Atomic Collisions: A Symposium in Honor of Christopher Bottcher (1945-1993) (Oak Ridge, TN 1994)	95-78689	1-56396-322-1
No. 350	International Symposium on Vector Boson Self-Interactions (Los Angeles, CA, 1995)	95-79865	1-56396-520-8
No. 351	The Physics of Beams Andrew Sessler Symposium (Los Angeles, CA, 1993)	95-80479	1-56396-376-0
No. 353	13th NREL Photovoltaic Program Review (Lakewood, CO, 1995)	95-80662	1-56396-510-0